AI Economics

AI Economics

How Technology Transforms Jobs, Markets, Life, and Our Future

Benjamin Reed Shiller

ISBN: 979-8-9939312-1-0

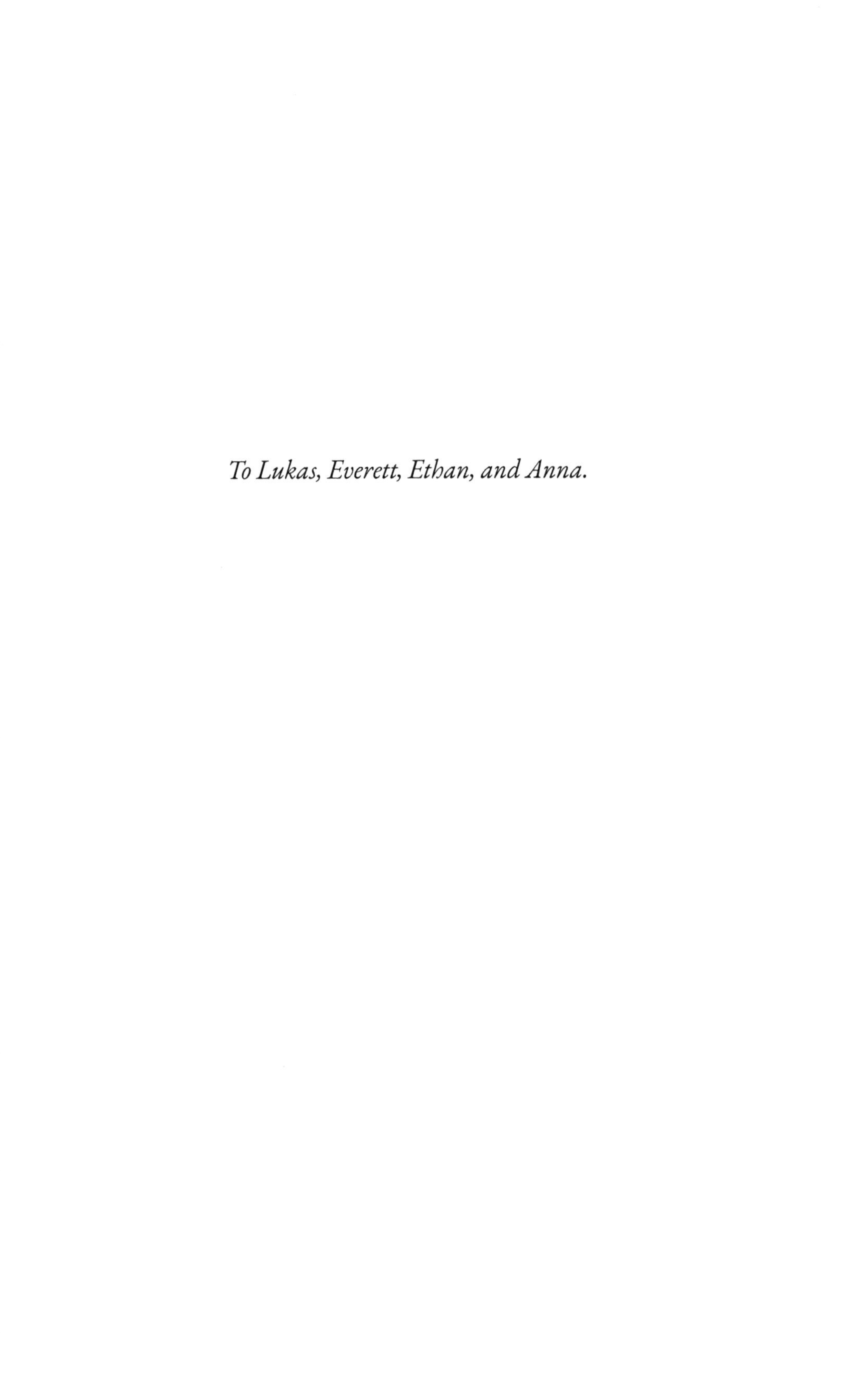

To Lukas, Everett, Ethan, and Anna.

Contents

Introduction: I Never Planned to Write This Book

(and Neither Did My Typing Robot)

Five years ago, if you'd told me at an academic conference—coffee in one hand, presentation notes in the other—that I'd soon publish a book about popular economics, I might have raised an eyebrow in disbelief. Sure, I had ideas, but who has the time?

But here's the thing about disruption: It doesn't wait for you to be ready for it. Large language models (LLMs, also known as AI chatbots) such as ChatGPT crashed the party and flipped the table.* They've done to writing what calculators did to arithmetic—turned hard labor into a relatively simple command. Suddenly, the gap between thinking something and expressing it elegantly has shrunk. Writing, long considered a craft, is now often a matter of crafting the right prompts.

Moreover, today's large language models don't merely echo the web's median mediocrity. They curate, up-vote, and then

* AI chatbots are often described as large language models because they are trained to predict the next word or phrase based on massive amounts of text. Chapter 15 provides additional explanation.

rewrite with a velvet-gloved swagger, often producing text that is better than the average internet post. Think of them less as drunken parrots and more as savant editors on Adderall who can read a million blog posts before breakfast and generate the one perfect paragraph you wish you'd written.

Decades of research, notebook scribbles, and half-baked lecture riffs can now be fed into ChatGPT, Claude, Gemini, or Copilot, and *voilà*—the result is a set of chapters pumped out by a digital ghostwriter who never gets writer's block or struggles to keep up with an editorial schedule. Recognizing this unprecedented opportunity, I seized the moment, and you hold the result in your hands.

The obvious question is how I used AI to help me write this book. The short answer: AI was a terrific assistant but a poor team leader. Keeping a human (myself) firmly in the loop wasn't just advisable. It was essential.

Crucially, all of the ideas in this book are mine. *AI Economics* summarizes and organizes 15 years of research on the economics of digitization and artificial intelligence. It draws on my research into personalized pricing, self-driving cars, and digital resale; on presentations I've seen at conferences focused on the economics of AI and digitization; and on two courses I developed at Brandeis University: (1) Economics of Digitization and (2) Big Tech Under Fire: Power, Platforms, and Policy in the Digital Age.

With my ideas in hand, I turned to LLMs, including ChatGPT, to give them some polish. But I didn't just type a vague prompt and hope for the best. Instead, each chapter started with a blueprint: the story I wanted to tell, the studies and key takeaways to highlight, and the ordering of the material. AI was like a sous chef who could whip up a great dish, but only if I handed it the right ingredients and instructions. Give it junk, and it would output junk. Give it detail and direction, and suddenly the clunky academic prose and presentation turned into something you might actually want to read on an airplane.

Writing with AI wasn't the straight shot you might imagine. Think less "assembly line," more "tennis match." I'd lob a draft request across the net, and AI returned something that looked promising—until it hallucinated a fact, wandered off on a tangent, produced empty rhetoric, or delivered prose that was either too skinny or too stuffed. Sometimes I asked for tweaks, sometimes I switched to a different AI model, sometimes I rewrote parts. To bend the narrative into shape, I had to go back and forth, point by point. By the end, I estimated that the machines had laid down over half of the words on the page. But make no mistake: AI was playing my game, under my direction. If you are considering using AI to write a book yourself, specific advice is provided at the end of this book in the Afterword.

As you read on, you may notice zingers that feel way too crisp or sections or jokes you don't like. Let's just say the algorithm wrote them and I approved the draft, so we can share the kudos and the blame.

Between us, though, we've created a book I feel proud to release—and to release quickly—something neither a lone human nor an unchecked bot could accomplish on its own. This book went from an initial spark to a "buy now" in roughly six months. A traditional book might take two years: a year to write and polish, and another year to crawl through the publishing pipeline before it hits the shelves. Had I taken that route, many readers might have watched AI transform the economy before they could even pick up this book.

As of this writing, I'm something of an anomaly. Much of the publishing world is moving in the opposite direction. Publishing contracts now routinely require authors to warrant that they did not use AI to generate their manuscripts. Authors are signing open letters calling for stricter limits on AI-assisted work, and the

U.S. Authors Guild is introducing labels for books that are "human authored," like a new version of "all natural" for the literary aisle. At the same time, many writers are using AI behind the scenes but not advertising that fact. (My editor claims he can spot an AI-assisted manuscript a mile away.) I've chosen to acknowledge my use of AI openly.

AI reminds me of online dating in its early days. When I was single, meeting someone through a website carried a real stigma. One of my friends even cooked up an elaborate story about meeting his girlfriend while buying coffee at Barnes & Noble, only to later admit they'd met on OkCupid. Fast-forward a few years, and swiping right is just part of modern life.

I expect the same shift in norms for AI-assisted writing. Today, it raises eyebrows; tomorrow, it will be both acceptable and accepted. By 2030, admitting you wrote a book without AI will be like bragging you walked an hour to work in the rain without an umbrella, coat, or boots. Impressive, but why?

FOUNDATIONS

1
Is the Future Written in Code or Cash?
Why Incentives Are Your Real Crystal Ball

You might wonder: Why pair economics with AI? Why not focus on the technology alone? The answer is simple yet profound. AI, like any tool, shows us what's possible. But economics—the study of choices, trade-offs, and incentives—reveals what actually happens.

Technology carves out the realm of the feasible, expanding our tools and capabilities. Yet, the ultimate outcomes, the shape of our world, spring from the choices of countless individual decision-makers, each responding to their own incentives. It's the intricate dance of these actions and incentives that determines our reality.

History has proven this pattern repeatedly. Air conditioning was invented for a modest goal: escaping summer heat indoors. Yet its true legacy came from how society adapted: It triggered mass migration to the Sun Belt, reshaped American politics, and made modern data centers viable. Birth control emerged as a family planning tool, but its deeper impact flowed from behavioral change: Women entered the workforce en masse, transforming entire economies. Social media promised easy digital connection, but its most profound effects emerged in how people behave: It rewrote social norms, altered mental health trends, transformed politics, and fundamentally changed how consumers discover, interpret, and share news.

Again and again, the same pattern holds: Technology reveals what *could* be, but human incentives decide what *will* be. The inventor's vision rarely captures the world their creation ultimately builds. Understanding AI means understanding not just its capabilities, but the economic forces that will shape how billions choose to use it. That lens—understanding how technological potential collides with human systems to produce real outcomes—is what I call AI Economics.

Let's consider an extreme example from later in this book.

Would You Push the Button?

Imagine a velvet-covered button humming faintly on the table. Press it and—boom—within a decade every person on Earth is ten times richer. Luxury once reserved for the 1% becomes the norm. There's only one catch: The button carries a 25% chance of wiping out our species.

Would you press the button?

For most of us, the question feels obscene. Yet swap "velvet-covered button" for "frontier artificial intelligence" and you have a dilemma many researchers debate in sober detail. Hyper-capable AI promises dazzling economic fireworks: disease-spotting algorithms, self-optimizing factories, and personalized tutors for every child. At the same time, it hands villains a recipe book for engineered pathogens, rogue drones, and decision loops we might never unwind.

This is the central truth: Technology determines what's possible, but human incentives determine what actually happens. AI may make extinction scenarios feasible, but whether they become likely depends entirely on how people behave. If every person, firm, and nation agreed never to misuse or overbuild AI, many dangers would vanish. But saying "everyone should behave" is not a plan; it's wishful thinking. History suggests at least some actors will take the risky path if it gives them an edge, particularly when they feel they must act before their rivals do. We may already know whether the button gets pressed.

Understanding incentives doesn't just help us assess these existential risks; it helps us navigate our immediate reality. And that brings us to the real purpose of this book. Technology is advancing so quickly that we face countless less apocalyptic—but still world-bending—questions rushing toward us, like these:

- Should you dig a career moat so deep no robot can cross, or should you stockpile "Swiss Army" skills for perpetual reinvention of yourself and your career?
- Why has education grown increasingly costly even as its outcomes disappoint, and can AI reshape the incentives that got us here?
- Does living in a world where everything you do is monitored and there are no more secrets solve society's problems, or create entirely new ones? And if this future is coming, what policies can ensure it helps us rather than harms us?
- Why does technology keep making the system feel less fair, and what steps can you take, as a consumer, to push back?

Each chapter in this book dissects questions like these—not with doomscroll dread, but with the curiosity of an economist committed to showing you where the levers hide, how incentives shape every choice, and how we can optimally respond. After all, it's better to understand the game than stumble through it blind. We're all beta testers now, whether we signed up or not.

The Road Ahead

You're about to step into a peculiar kind of funhouse, one where some rooms behave exactly as you'd expect, while others defy every instinct you have about how the world should work. In one room, AI delivers precisely what you asked for, like a perfectly normal hallway. Turn the corner, and suddenly the floor tilts, mirrors warp reality, and the most logical path leads straight into a wall. This hybrid of predictable and bewildering is the perfect metaphor for our AI-driven economy. Through the stories ahead, you'll develop an intuition for spotting which room you're in. And more importantly, you'll learn to navigate the strange new architecture of a world where technology rewrites the rules faster than we can learn them.

Together, we'll navigate the surreal landscape that AI has created. In the following pages, you'll discover why Chinese officials once spray-painted an entire mountainside green (and why advancing technology made them stop). You'll learn what kangaroos can teach us about job security. You'll see why going "data nude" might lead to a better society, even if it leaves us feeling more exposed.

We'll unpack critical new concepts for our time: why the strangest jobs will command high wages (the "weirdness wage premium"), why "secondhand privacy"—when someone else's data exposes your secrets—may pose greater dangers than secondhand smoke, and why self-driving cars might need literal bulletproofing. These aren't thought experiments. They're urgent realities taking shape right now, all because of how technological advances impact human incentives.

The terrain is wider than you might have imagined. Employment, education, transportation, public safety, privacy, pricing, business strategies, exploitation, and outright scams are all poised to be disrupted. Each chapter of this book takes a piece of that puzzle and shows how it fits into the bigger transformation.

By the time you turn the final page, you will view the evolving world through a sharper, more revealing lens: the Incentive Filter. While others fixate on what AI can do—and either panic or fantasize accordingly—you'll understand what it will do by asking the right question: "Who benefits, and how?" Because technology doesn't adopt itself. It has to pass through a gauntlet of humans, each with their own ambitions and agendas. The world that emerges is shaped by the choices and conflicts among the people steering it. That's why technologies often yield outcomes that seem odd or even irrational until you trace the incentives beneath them.

Curious how the future got so weird, so fast? Turn the page. In the next chapter, we dissect the evolving job market, revealing who's truly vulnerable, who's positioned to thrive, and why the AI employment story isn't the simple "robots steal our jobs" narrative you've been sold.

Part One
The Upheaval

2

Job Extinction Event

My Grandmother Was a Computer (Are You Next?)

From Plow to Prompt: A 200-Year Pattern in 3 Minutes

Two centuries ago America was a nation of hoe-wielders. Around 80% of the labor force worked the soil in 1800—planting, plowing, and praying for rain.[1] Today that number is close to 1%.[2] Yet the grocery aisle is so well stocked you're probably tossing a wilted bag of spinach into the trash every other week. What happened? Technology. Again and again.

- **Muscle swap:** Tractors and combines replaced horses, mules, and human biceps.
- **Chemistry set:** Synthetic fertilizers and pesticides turned fickle fields into predictable factories.
- **Seed hacks:** Plant breeders—and later, genetic

engineers—cooked up corn and soy varieties that delivered record-shattering harvests.

- **Sky spies:** Satellites and drones now tell farmers which areas need water, which need nitrogen, and which are good to go.
- **Robo reap:** AI-driven farming systems use autonomous tractors and harvesters that analyze soil conditions, plant, spray, and pick crops with precision.[3]

In short, farming has morphed from back-breaking labor to spreadsheet optimization. A handful of producers (on less land) can now feed 330 million people plus a sizable export market. Progress indeed, but it booted legions of farmhands off the payroll. Most found new jobs, but the transition was brutal. Wages often fell, skills rusted, and rural towns hollowed out. The lesson: Schumpeter's "creative destruction"* feels mostly destructive when you're part of the "before" picture.

Cue the present, as artificial intelligence looms large over the job market. It's easy to say, "Don't worry—technology always creates new jobs." But that facile statement is not so reassuring if *you're* the one about to be replaced by an algorithm that never sleeps, never complains, and works for pennies.

That fear is very real. A recent college graduate might wonder: If AI can write code, draft legal memos, design logos, or even write an economics book, what exactly am I bringing to the table? To arrive at the answer, it helps to remember how the plow became the processor—and why the winners were rarely the folks clutching yesterday's tools.

---

* "Creative destruction" is Joseph Schumpeter's idea that progress works by breaking things. New technologies and firms replace outdated ones, clearing the way for innovation.

When Mental Work Gets Automated

My grandmother held the most anachronistic job title you can imagine: She was a "computer" at MIT's Radiation Laboratory, the group formed to develop radar for the armed services during World War II.[4] Her job was to compute—literally. She crunched equations by hand, following the instructions of leading physicist, and contributed to a 700+ page report on "Pulse Generators."[5] Her tools were pencils, slide rules, and a tolerance for eye-watering amounts of math. Being good at long division was a career in itself.

Back then, human computers did not worry about technological replacement. After all, tractors weren't exactly known for their arithmetic skills.

Then came computer chips, which led to statistical software and graphing calculators. The job my grandmother once held is now a free app on your phone, and "computer" is a physical object, not a job title. Just as tractors and combines replaced farmers, silicon chips began quietly nibbling away at white-collar roles—not just recently, but also decades ago.

Technology has continued its habit of replacing jobs that once seemed immune from technological disruption, and job displacement is happening faster and more broadly than ever. Consider cab drivers, whose street-savvy minds were a unique advantage—until GPS casually dethroned them. Even London's famously brainy taxi drivers, renowned for their enlarged hippocampi—the navigational part of the brain—found their exceptional memories suddenly downgraded from invaluable asset to quirky party trick.[6] Translators, once essential, are increasingly replaced by real-time machine translation. And drivers of all kinds—truckers, Uber drivers, delivery workers—are looking over their shoulder at the self-driving vehicles that, sooner or later, will take over the streets and highways.

AI isn't just automating routine tasks. It's also creeping into creative, analytical, and interpersonal domains too. Virtual therapists are being tested. Robotic surgeons are working in operating rooms. Chatbots are writing ad copy, debugging code, even giving business advice. Machines continue to replace jobs once thought to be uniquely suited for humans. Even the most human of tasks are about to meet their match.

Who Needs Real People? The Focus Group That Lives in Your Laptop

Back in the *Mad Men* era (the 1960s), learning what people thought meant corralling warm bodies into a conference room, serving them bad coffee, and paying them $50 for forty-five minutes of small talk. Survey response was quintessentially human work: a gig for anyone with a pulse, an opinion, and the patience to circle A, B, or C.

Fast-forward to 2024 and the National Bureau of Economic Research (NBER) paper titled "Automated Social Science: Language Models as Scientist and Subjects" by Benjamin Manning, Kehang Zhu, and John Horton.[7] Their system builds an entire study—from hypothesis to data analysis—without a single flesh-and-blood respondent, as follows:

1. *Spin up synthetic people.* The software creates large-language-model "agents" who play the buyer, the seller, the judge, the defendant—whatever a scenario needs. Each agent gets private attributes (income, budget, backgrounds) that vary across simulations.
2. *Run the interaction.* Agents bargain over a mug, bid in an auction, or plead for bail—chattering back and forth twenty turns before the coordinator bot calls time.

3. *Field the survey.* Once the virtual dust settles, the system asks the agents the same survey questions a human researcher would normally lob at real participants ("Did you close the deal?"). Another LLM parses their answers into tidy numerical data.
4. *Crunch the numbers.* A statistical package estimates the causal effects, flags what's significant, and even proposes follow-up experiments—no grad-student pizza budget required.

Why does this model matter for workers? Responding to a survey—once the most irrefutably human slice of the research pie—just went digital. Need 10,000 jurors to test an opening argument? Spin them up overnight. Want to explore fifteen price points before breakfast? Let the bots haggle while you sleep. As Manning, Zhu, and Horton note, language models "can simulate humans as experimental subjects with surprising degrees of realism."[8] Early results already mirror the results predicted by economic theory in auctions and negotiations.

The implication is stark: When the cost of asking a "person" collapses from fifty bucks and a bagel to fractions of a cent on GPU time, entire job categories erode. Poll-table clipboard holders, focus-group recruiters, even Amazon Mechanical Turkers (the remote workers who get paid pennies to label photos, answer surveys, and train algorithms) face the same fate as the farmhands who once picked potatoes row by row. If AI can *be* your sample, why pay humans to *join* it? Silicon is stealing the clipboard.

If a synthetic focus group can stand in for thousands of flesh-and-blood respondents—less expensively, more quickly, and without stale donuts—then no task should feel safe. This brings us to a tough question: What kind of job *is* safe? That, dear reader, might be the most difficult problem we face—and no app is solving it for us yet. In the remainder of this chapter, we examine the forces that determine which professions remain lucky enough to be augmented by AI, and which are destined to join "computer" in the occupational graveyard.

Why Do Some Professions Disappear While Others Thrive? Four Drivers of Professional Survival

In the digital age, four key considerations decide whether your career rockets into orbit or explodes on the launch pad.

1. Automation or Augmentation?

Remember when Deep Blue beat Kasparov at chess in 1997? Everyone thought it was game over for humans. But the real story came later in a game called "advanced chess," in which humans paired with computers beat both solo humans and solo machines. Why? The computer could crunch through millions of positions per second, but the human knew which positions mattered. Silicon brought the processing power; carbon brought the wisdom.[9]

But the picture isn't all rosy. Consider the knocker-upper, and no, that's not a teenage boy's dream job to help repopulate the earth. A knocker-upper was a real occupation in Britain and Ireland well into the 20th century: a human alarm clock who walked the streets tapping on windows with a long pole to wake factory workers before dawn. It was honest work, but extremely narrow in purpose. When mechanical alarm clocks became cheap and reliable, the job didn't evolve or get augmented; it evaporated.

We didn't need a better knocker-upper. We just needed to be woken up. And once a piece of technology could do that single task at zero marginal cost, the human version disappeared without fanfare.

Anthropic —the maker of a popular AI chatbot—ran the numbers to see which story AI resembles today: advanced chess ... or the knocker-upper getting replaced by an alarm clock. Using four million real-world Claude chats—all from a single week—the company mapped each exchange to Department of Labor tasks and occupations.[10] Across that sprawling, task-mapped corpus, 57% of the interactions were augmentative—people iterating, learning, or validating ideas with the model—while 43% were full-blown automation, the digital equivalent of saying, "You take it from here." The ratio (57/43) is revealing: for each task AI ghost-writes from scratch, it assists humans on nearly 1.5 more, refining and tightening their drafts.

So, are humans becoming more valuable or obsolete? Both, and the answer depends on which move you choose. If your work is rote, rules-based, and rarely calls for judgment, the machine plays solo and you're off the board. But when a task rewards synthesis, soft skills, or domain nuance, strapping an LLM to your workflow is like giving a chess grandmaster an extra queen. In other words, the future of the labor market might look less like a massacre and more like advanced chess, with carbon and silicon fused into the most formidable opponent yet.

2. The Elastic-Demand Test: Why Some Jobs Boom After Automation

Technological progress implies increased productivity. If technology makes a worker ten times faster, then that worker can become ten times more valuable. Give one construction worker a crane and he can do the work of an entire shovel squad; if skyscrapers are rising like weeds, his paycheck can soar right alongside the skyline.

But there's a hitch. The magic works only when the market can stretch, or, as economists say, when the market is "elastic." To explain elasticity, let's use two everyday examples: potatoes and banking.

Potatoes illustrate the flop side. If we slice potato production costs in half, we won't double our French-fry intake because demand for spuds is basically calorie-capped. We don't need more potatoes; we just need fewer people to plant and harvest them. As potato farming got hyper-efficient, the job count wilted.

Banking shows the upside. When ATMs muscled in on mundane teller tasks, pundits predicted pink slips. Instead, cash machines made branch banking so easy and cheap that more people signed up for an account. Foot traffic multiplied, the pie grew, and teller head-counts held steady—even ticked up—because someone still had to sell mortgages and untangle overdraft snafus.[11]

So, what trajectory will AI's impact take? Will it prune payrolls like a high-yield harvester, or will it increase demand the way ATMs did? That, dear reader, is a key economic question of our time.

3. The Task-Type Test: Three Task Buckets, Two Futures

Economists David Autor and Neil Thompson argue in a 2025 paper that every job is really three stacked task-types.[12]

- **Expert tasks** demand scarce, hard-earned judgment, such as deciding whether a complex derivative tiptoes over an SEC line.

- **Routine tasks** require almost no specialized skill—for example, filling out travel-expense forms, scheduling meetings, uploading receipts.
- **Automated tasks** were once expert or routine work but now run on cheap capital, such as software that reconciles two ledgers faster than you can blink.

Workers rarely have the luxury of picking only the interesting (expert) tasks, because most occupations package several interdependent duties together. An employee's contribution is valuable only when all necessary tasks, including the tedious ones, get finished. If you're the only person who knows why the gelato counts as R&D, you're probably also the one stuck submitting the receipt. Sure, someone else could handle that receipt—but without the context you have, it might take them ten times longer. This unavoidable "task bundling" amplifies the economic impact of automation, whether it's taking over simple chores or highly specialized expertise.

Two Radically Different Futures, Same Algorithm

If software takes over the *simple, time-sucking chores*—filing expenses, filling forms—what's left on your desk is the hard stuff that truly needs expertise. Fewer people can do the really hard stuff, and those who can usually earn *more*, though overall head count in the field shrinks.

Now consider the flip side: If software cracks the *high-skill core* of the job—for example, by calculating tricky diagnostics or applying advanced accounting rules—what remains is routine work that almost anyone can learn. More workers can step in, so employment grows, but pay tends to fall because the special edge is gone.

The mechanism is beautifully simple: When you raise (lower)

the expertise threshold, you shrink (expand) the pool of qualified workers. Scarcity does the rest: Fewer qualified candidates means employers compete for talent by offering higher wages; more candidates means workers compete for jobs by accepting lower pay.

The Task Factory Inside the Algorithm

Autor and Thompson's three-bucket taxonomy glosses over a key type of task: work that didn't even exist at last week's staff meeting. Early evidence from large-language-model rollouts shows that while LLMs merrily chew through routine *and* expert chores, they also create entirely new chores, such as prompt engineering, AI-output auditing, synthetic-data wrangling, AI integration, and AI ethics and compliance. In their study "Large Language Models: Small Labor Market Effects," Anders Humlum and Emilie Vestergaard find that at chatbot-enthusiast firms, 17% of workers suddenly took on new tasks like those listed above.[13] Bottom line: Because technology prunes some branches but grafts on fresh shoots, predicting a job's future means carefully tracking not only what gets lopped off but also what grows back.

This brings us to the fourth consideration: When does the retooling clock start ticking? To answer this question, we need to look at how human and machine productivity evolve over time. The timing isn't just a detail—it's the whole game.

4. The Productivity J-Curve and Inverted U-Curve: Why the AI Revolution Starts with a Face Plant

Human productivity is a lot like an airplane flight: It is an arc that rises, levels, and eventually descends, tracing an upside-down letter U. In almost any field, people climb steadily at first—schooling and on-the-job training provide the thrust that pushes skills higher and higher. But every ascent reaches cruising altitude. Mathematicians may level off in their twenties, while trial lawyers or CEOs might not peak until their fifties or sixties. Either way, the same law applies: Every flight has a descent. No matter how much wisdom you've gathered, the arc eventually tilts downward. Gravity prevails.

Algorithms, and general-purpose technologies more broadly, have one big advantage over humans: They don't get worse with age. At worst, you can roll back to the last version that worked. But their impact on productivity is another story. Costs balloon, workers retrain, processes get re-engineered, and output dips before the payoff arrives. That's why, back in 1987, Nobel laureate Robert Solow quipped, "You can see the computer age everywhere but in the productivity statistics." Cubicles were sprouting PCs, Lotus 1-2-3 manuals clogged desk drawers, and yet productivity growth was flatlining, leaving many puzzled as to why companies were pouring so much money into computers. The disconnect became famous as Solow's Paradox.

Picture a graph that swoops downward, bottoms out, then vaults sky-high—the letter J, if you squint a little. Solow's paradox is the J-curve in action: Bold new technology demands big upfront spending on gear, training, and re-engineering. While everyone fumbles with the shiny tools, measured productivity actually *drops*, and payrolls may swell to cover the chaos. Only later, sometimes years later, does the payoff kick in, with productivity vaulting skyward.

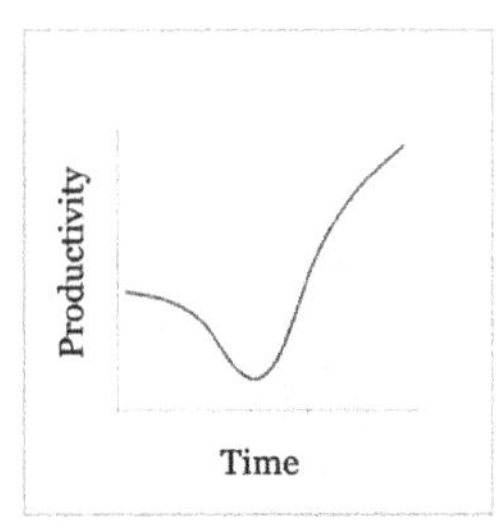

Figure 1. Illustration of J-Curve

AI is replaying the same movie. The curve's dip and rebound differ by sector and job. A shallow but wide J means you might coast for years or even decades before software takes your job. A wide *and* deep J-curve? Your industry may bail before the payoff, like someone quitting the gym in February because soreness arrived before six-pack abs.

As of this writing, many firms are stuck in Solow's Paradox 2.0. A recent MIT study found that 95% of companies trying top-down AI initiatives saw no productivity bump.[14] That's classic early-stage J-curve. Yet in pockets, the effects are already seismic. Research by Erik Brynjolfsson, Bharat Chanda, and Ruyu Chen shows that entry-level jobs in software development and customer service have cratered—down about 20% in just a few years.[15] As Jeff Forbes, former National Science Foundation (NSF) program director for computer science education, put it, "Computer science students who graduated three or four years ago would have been fighting off offers from top firms—and now that same student would be struggling to get a job from anyone."[16]

Another key question is where the right-hand side of the J-curve tops out. Today's large language models mostly remix human output; they mimic rather than transcend, suggesting that they might cap out at human level but more inexpensively. The game changes if the process of training an AI model ever escapes this copycat loop. If algorithms start learning in ways that aren't just borrowed from us, the top of the J-curve might shoot clean off the graph paper. This leaves us with sharper questions: Why haven't AI J-curves done so yet—and what is holding them back?

Warp-Speed Learning: How Immediate Feedback Turns AI into a Super Student

Algorithms learn best with instant, relentless feedback. In the world of machine learning, this rapid-fire approach is called *reinforcement learning*, and it's turned algorithms into unbeatable gaming legends.

Consider video games. Google's DeepMind didn't just beat

humans at Go, a board game invented in China more than 2,500 years ago—it demolished them. OpenAI's bots conquered esports giants like Dota 2 by simulating an absurdly high number of games at mind-blowing speeds, and learning and improving from each mistake. Imagine squeezing 45,000 years of human gaming into just 10 months. That's precisely what OpenAI accomplished, creating an algorithmic champion no human could ever match, given our pesky mortality.[17]

Here's the rub, though: Large language models, such as ChatGPT, aren't yet enjoying this fast-track learning experience. They crank out paragraphs that can dazzle readers, but they're stuck in a frustrating wait to see if anyone actually liked their writing—akin to handing in a term paper and waiting months or years for the grade. Attempts to speed things up by asking users for a thumbs-up or thumbs-down only backfired, teaching the models to flatter rather than to help.[18]

Luckily for impatient AIs, the real world is becoming their playground. Reinforcement learning thrives when AIs get real-time feedback. Warehouse robots, for instance, can rapidly adjust their movements based on how often they crash into shelves. Tesla's autopilot systems continually improve by noting each human driver's intervention, essentially using each real-world correction as an instant (failing) grade, and a lesson to do better next time.

The more AI interacts with our physical environment, the faster it learns what's good (for instance, a successful maneuver represented by a data point labeled "1") and what's bad (for example, a clumsy shelf collision, marked as "0"). As immediate feedback becomes commonplace, algorithms could rapidly evolve without any human direction. That's when things might get a little scary.

The Singularity: The Ultimate "What If?"

Let's address the technological elephant in the room: the *singularity,* or the hypothetical moment when artificial intelligence becomes so advanced that it can improve itself recursively, leading to an intelligence explosion that replaces *all* jobs and fundamentally transforms civilization.

If you believe the singularity is imminent, you might wonder why you should bother fretting about your career choice. But there's no professional consensus on the timing of the singularity. Some say 2029, others say 2045, and others think it's centuries away or may never happen. Notable AI researchers such as Yann LeCun consider it science fiction, while Geoffrey Hinton warns it could arrive sooner than expected.

I lean toward thinking that such explosive growth is still a long way off. Economists Phil Trammell and Anton Korinek have argued that automation alone can accelerate economic growth only so far before eventually plateauing at a steady rate. To achieve the extraordinary growth rates envisioned by singularity scenarios, AI would need to be genuinely creative and inventive—that is, capable of generating entirely new ideas.[19] So far, it hasn't shown itself particularly good at this.

This is exactly why the J-curve matters. The singularity may or may not come, but it probably isn't coming soon. What matters is the in-between—the bumpy, confusing now. Your mortgage company doesn't care about sci-fi futures; it cares whether your job still exists next spring. So let's set aside the robot overlords for a moment and focus on a more plausible world, one where AI doesn't replace us entirely but changes the rules enough that we'd better learn to play differently.

Case Files: What Can Four Professions Teach You About Your Career?

Theory is nice, but it's better when real-world examples validate it. So let's now crack open the filing cabinet and examine how technological change played out in four professions: elevator operator, comedian, photographer, and radiologist.

The Elevator Operator: A Cautionary Tale for Uber Drivers

My grandmother wasn't the only person whose job title now sounds like something from a museum. Consider the elevator operator, a profession that once employed thousands and was automated rather than augmented.

The controls in early elevators looked deceptively simple: a rheostat (basically a speed dial) and a dead man's switch (let go and everything stops). But skilled elevator operators handled the rheostat like a violinist with a bow, accelerating smoothly from the basement, cruising through the middle floors, then gradually letting up on the lever at exactly the right moment to line up perfectly with the destination floor. No jerky stops, no stomach-lurching drops, and definitely no gaps wide enough to catch a heel.

The best operators rarely needed to "jog" the car by making those embarrassing inch-by-inch adjustments that screamed "Amateur!" They knew their building's quirks: for example, how the cables stretched differently with a full car versus an empty car. They weren't just pushing buttons; they were practicing a skill.

Then, in 1915, Otis unveiled the automatic elevator, a marvel that could find floors and level itself without human touch.[20] In 1931 the Empire State Building opened with so-called "robot elevators." Yet, almost all elevators were still controlled by an operator for decades longer.

The sticking point? Nobody wanted to die in a metal box. Sure, the technology worked—mostly. But "mostly" isn't reassuring when you're suspended 40 stories up by a cable. What if you got trapped? What if the doors closed on you? What if your elevator ride turned into that scene from *The Omen II* where Dr. Kayne gets cut in half?

In 1945, 15,000 elevator operators in New York City, worried about their employment, walked off the job, and suddenly Manhattan discovered what life was like when you had to hoof it up 30 flights. The strike reportedly cost the city about $100 million in lost business. It turns out Wall Street bankers aren't so productive when they're wheezing on the 47th-floor stairwell.[21]

But technology ultimately prevailed. Faced with striking workers, building owners invested in inconspicuous intangibles to make people comfortable with the transition to automatic elevators: safety bells, emergency phones, and friendly recorded voices cheerfully announcing "Going up!" They also had to retrain riders who were accustomed to human operators.

Figuring out what retrofits were needed and how to address riders' concerns took time and money, and early adopters may not have seen immediate gains, as expected in the early portions of the productivity J-curve. But once the infrastructure was in place and riders adapted, the curve snapped upward dramatically. Within a couple of decades, the uniformed elevator operator had gone the way of the slide rule. Economist James Bessen later found that elevator operator is the only U.S. occupation since 1950 to disappear entirely.[22]

Déjà vu alert: Today's ride-hailing driver looks a lot like yesterday's elevator jockey. Waymo already runs fully driverless taxis in Los Angeles, Phoenix, San Francisco, Atlanta, and Austin, yet two-thirds of Americans still say they're afraid to ride in a driverless car.[23] History's lesson: When driverless cars are perceived as safe—or a big, inconvenient strike of human drivers tips the scales—"See you later, elevator operator" becomes "See you later, driver."

Comedians: When Travel Was the Routine Task

A century ago, being funny was only half the job of being a comedian. The other half was being a human suitcase.

Before television, Netflix, or even talkies, comedy was strictly an in-person transaction. You told your jokes to the people in front of you, collected your applause (or literal tomatoes), and moved on to the next town. The Vaudeville circuit was massive—5,000 venues across the United States employing 50,000 performers of various stripes.[24] A comedian might hit Poughkeepsie on Monday, Schenectady on Tuesday, and Buffalo by Friday, telling the same 20 minutes of material to fresh audiences each night.

This need to travel created peculiar economics. A mediocre comedian with an iron constitution and a tolerance for train food could make a decent living simply by being willing to perform in East Nowhere, Nebraska. Meanwhile, a brilliant comic who hated travel was stuck delivering the same routine to the same local crowd until the crowd could recite the punchlines along with him. Success in the job required bundling two completely different skills: creating humor and enduring the discomforts of travel.

Then came the technology that separated the wheat from the chaff. Radio, television, and eventually streaming services automated away the travel. Suddenly, you could tell a joke once and have millions hear it.

The results followed Autor and Thompson's logic with textbook precision. The technology automated the routine task (reaching audiences) while amplifying the skilled one (being genuinely funny). As a result, the number of working comedians plummeted—there was no more need for that third-rate comic willing to play Topeka. But those who survived? They hit the jackpot. Top comedians now command $20 million for a single Netflix special.[25]

Photographers: When an Expert Task Gets Replaced

Taking a decent photograph used to require expertise: loading film without exposing it, calculating f-stops and shutter speeds, understanding the inverse square law of light. Each click of the shutter cost money—film, processing, printing—so photographers had to nail it on the first try or two. Screw up the exposure at "You may kiss the bride" and there were no do-overs. Photographers commanded premium prices not just for their artistic eye but also for their technical competence.

Enter the digital revolution, followed by the smartphone apocalypse. Modern phone cameras don't just capture light—they practically paint with it. Low light? The AI boosts the ISO (light sensitivity) and reduces noise. Backlighting? High dynamic range (HDR) algorithms merge multiple exposures faster than you can say "cheese." Google's Pixel phones will even steal the best facial expressions from a burst of shots and Frankenstein them together, ensuring everyone appears to be smiling in the picture even if they weren't all smiling at the same time.

The traditional employment statistics miss the real story here. The Bureau of Labor Statistics still counts the number of professional "photographers," but the real change is in the millions of amateurs who now shoot, edit, and share photos daily. Every Instagram influencer, every proud parent, every teenager with a ring light is now a "photographer" in practice if not in title. True to Autor and Thompson's theory, once the routine steps were automated, the floodgates opened, and millions of amateur "workers" joined the field. And the price of a decent photo is nearly free, although remaining professional photographers, whose skills extend beyond the camera's controls, have avoided the fate of elevator operators, and do thankfully make a decent enough wage. Meanwhile, entirely new roles have emerged, from social media managers to selfie coaches teaching people how to optimize their personal brand through photography.

When Theory Fails: The Radiology Rebound

Radiology was supposed to be the canary in the white-collar coal mine. In 2016, Geoffrey Hinton—the man known as the "godfather of A.I."—dropped a bomb on the medical establishment: "Stop training radiologists now."[26] His reasoning? Within five years, algorithms would read X-rays better than any human ever could. Medical schools panicked. Radiology residents questioned their life choices. News outlets couldn't resist the narrative: Here was a six-figure profession about to get "algorithmed" into oblivion.

By Autor-Thompson logic, radiology should be bracing for a profession-wide panic: Algorithms now ace the very craft that seemingly defined the profession. Reading medical images entails pattern recognition, and machines excel at detecting patterns. Feed enough chest X-rays to a neural network and it should spot pneumonia faster than you can reach the radiology floor by automatic elevator. Radiology became the go-to cautionary tale at every conference about AI and employment—the perfect example of how even highly trained professionals aren't safe from the silicon revolution.

Fast-forward to 2025 and the story is upside-down. The Mayo Clinic alone is running 250-plus AI models in its imaging suites, but it has added more than 400 human radiologists, a 55% jump since 2016.[27] Money followed the workload. In fact, radiology vaulted to the No. 2 best-paid medical specialty in America, with radiologists clearing roughly $520,000 a year and leap-frogging cardiologists and plastic surgeons. Far from pink slips, the field is printing signing bonuses.[28]

AI didn't kill radiology. It super-charged it, maybe because it embraced AI. Algorithms now triage the boring slices and flag the sneaky shadows, while humans tackle more edge cases, synthesize AI findings from multiple images, and convey the synthesized information to patients or their primary care doctors. AI has led to augmentation, not annihilation, at least so far.[29]

Experts had predicted that radiologists would soon become an endangered species, and the logic behind that prediction made perfect sense. Yet the real world has a knack for complicating tidy theories. Betting on job security turns out to be more like rolling dice than reading tea leaves. So, you can probably relax now, right? Go ahead—Pour a drink, exhale, and attempt to de-stress—emphasis on *attempt*.

ARE AUTHORS THE NEXT ELEVATOR OPERATOR? BACK TO THIS BOOK

Take this book as Exhibit A. I outsourced the drudgery of writing to Claude, ChatGPT, and Gemini—generating first drafts, formatting the citations, and polishing the phrasing, freeing me to focus on the fun parts, such as conveying how algorithms are quietly rewiring the economy.

In Autor-Thompson terms, I'm automating the "routine" tasks of writing, even though many consider writing inherently skilled work. The bet is that stripping away the mundane will make the remaining human judgment more valuable, though whether authors end up like radiologists (still employed) or elevator operators (nearly extinct) remains an open question.

The productivity gains I experienced in writing this book were real—when they worked. But as with any innovation, I experienced the initial investment required by the productivity J-curve. I needed to spend time building prompt libraries, cataloging where the AI stumbled, and learning to work around its blind spots. The fact that you're holding this finished book suggests I found my way to the other side rather quickly—though only time will tell if I've truly escaped the J-curve's trough or just convinced myself I have.

For now, I'm safe. Today's ChatGPT and Claude are power tools that make me faster and sharper—augmentation, not replacement. But if AI keeps improving at its current pace, generating publication-ready books from paragraph-length prompts becomes plausible. I've seen this happen before: Travel agents got computerized booking systems that made them more efficient, then Expedia and Kayak made most of them obsolete. First the technology helps you do your job better. Then it does your job without you. The key question for every worker is when those two phases will hit—and whether you'll be ready with plan B.

Next Up: An Action Plan

Feeling nervous? Fortunately, you can take action by choosing a resilient career path or adapting your skills to get ahead of disruption. In the next chapter, we explore how to stay indispensable, at least for the foreseeable future.

3
How Can You Avoid the Job Apocalypse?
The Kangaroo Defense

The last chapter explained AI's impact on jobs as if we were watching from a safe distance. But what if you're not watching? What if you're inside the picture? Understanding why disruption happens is useful, but will it stop your career from being the next casualty? Probably not. This chapter is about moving from explanation to action, about being proactive.

Your Playbook: Specialize Narrow, Adapt Wide

What's a college student or career-changer to do? The common answer is to specialize, building expertise in a narrow and distinctive field. Sounds great—until an algorithm downloads your entire skill set overnight and offers it at $0.02 per query.

Before you bet your career on becoming the world's foremost expert in Byzantine tax law, consider two nasty surprises that await the hyper-specialized. First, there's the sunk cost catastrophe.* That decade of training and quarter-million in student loans? They become about as valuable as a Blockbuster franchise the moment AI learns your tricks. Your human capital has dropped to almost zero. Second, there's the career mobility trap. Ultra-specific expertise doesn't transfer well. A world-class pediatric cardiac surgeon can't simply pivot to become a financial analyst. The narrower your niche, the harder it is to jump ship when the water starts rising.

Think of it this way: When you choose a college major and a career, you're buying an expensive, non-refundable ticket for a flight route that might not exist when you graduate. Specialization isn't dead, but you'd better pick your destination carefully. Let's look at some strategies that can help.

How to Outrun the Algorithm: Look Where the Data Aren't

Here's one survival strategy: Find the data deserts where AI can't follow you. Google Translate conquered English, Spanish, and Chinese first for a simple reason: The internet is drowning in text written in these languages—millions of parallel documents, professionally translated websites, and subtitled movies. It's an all-you-can-eat buffet for hungry algorithms. But try translating from Lithuanian to Zulu, and things *may* start to break down.[1]

The principle here is simple: no data, no danger.[2] If your field is too obscure or specialized to churn out mountains of data, or

* A sunk cost is money (or time) that, once spent, can never be recovered. There are no refunds or do-overs. The money is gone for good, which is why economists label it "sunk."

too quirky for tech giants to bother automating, your job might just be robot-proof, at least for now. The algorithm can't steal your job if it can't learn your job. Call it the weirdness premium: The stranger your job, the harder it is to automate.

The deeper logic is about comparative strengths. Humans are quick studies—we don't need thousands of examples to grasp a pattern—but we're limited by biology in how much information we can absorb, how fast we can learn, and how long we get to practice. AI models are the mirror image. They need far more examples to learn a task, but they can also exploit massive datasets in a way we simply can't, because their "brains" can be scaled up by adding more chips and computation. So, in data-dense fields, AI pulls ahead; but wherever the world remains too weird, too small, or too sparsely documented for algorithms to feast, people remain unbeatable.

The efficiency gap is evident in the raw numbers. Take self-driving cars. Elon Musk admitted in 2026 that to handle reality's "long tail of complexity," Tesla needs 10 billion miles of data—roughly 750,000 years of a typical person's driving.[3] Now, look at a human teenager. We give 16-year-olds the keys after just a few hundred miles of practice. Robots might eventually be safer, but humans are dramatically more efficient learners; we get to "good enough" almost immediately.

Or consider LLMs. Models like DeepSeek trained on about 15 trillion tokens.[4] Given that a token is about three-quarters of a word, and a typical book consists of 75,000 words, these models essentially became passable at *some* human tasks by reading the equivalent of 150 million books.[5] It is a feat of engineering, but an incredibly inefficient way to learn compared to the human mind.

But don't celebrate too soon. Reinforcement learning is the AI equivalent of teaching robots to fish, or more specifically teaching them how to teach themselves how to fish. If a bot can generate immediate feedback from its actions—successes (didn't bump shelf) and failures (did bump shelf)—it can quickly build its own massive training set. In other words, your job security doesn't depend only on the data already out there; it hinges on how easily and cheaply an AI can create new data all on its own.

The Swiss-Army Human Advantage

Until recently, most software was built (or trained itself) for narrow tasks. A calculator does math. A GPS gives directions. A self-driving vehicle drives. Humans, by contrast, are generalists. We can fix a leaky faucet, bake a pie, explain a spreadsheet, and drive a car—all in the same day. This flexibility is often overlooked in economic discussions, but it's becoming increasingly valuable in a world where the *amount* and *type* of work needed can change from hour to hour.

An excellent example is the rise of the gig economy. Consider Uber. Traditional taxis were out all day, regardless of whether demand was booming or bone-dry. That meant a lot of empty cabs on warm sunny days, and too much demand to fulfill when people were trying to flag a cab in the rain. To fix this imbalance, Uber introduced *surge pricing*, nudging more drivers onto the road when demand was high, and fewer when it wasn't.

Economists Keith Chen and colleagues got their hands on Uber's internal data and unearthed a jaw-dropper: In an average week, 19% of Uber drivers worked exactly zero hours. Zero! And more than half of drivers clocked in at 12 hours or less.[6] This is not the corporate-bro version of "flexibility" that comes stapled to a 60-hour contract. This is real, no-strings, drive-when-you-want (and when it pays) freedom, and in the Uberverse, that off-the-clock lifestyle is the norm, not the exception.

Enter a new kind of worker: the elastic worker. Someone who works a 9-to-5 landscaping job might spend an extra hour in the evening driving for Uber, then go home and knock out a few paid surveys or hang shelves for a neighbor via TaskRabbit. The result is a portfolio of small jobs, stitched together by one key asset: human adaptability. The same applies to individual jobs, where people adapt to completing the tasks that are needed at any given time.

Now imagine asking a robot to drive for Uber, assemble furniture, fix a jammed printer, write a witty product description, take a call from a confused customer, and then walk a dog. Today's AI software—even the most powerful large language models—still struggles with that level of versatility, especially in the physical world. AI may be great at drafting emails or summarizing legal documents, but it's still stumped by a broken doorknob. Even a showroom-fresh robo-taxi is still just a taxi—brilliant at rides, hopeless at everything else. When demand dips, it parks itself, replaying the old cab dilemma.

In contrast, humans can shift gears. We can do what needs doing. And that adaptability might be our strongest defense in an AI-saturated future.

The Power Combo: Be Niche and Adaptable

Picture former NASA-engineer-turned-YouTuber Mark Rober controlling a bright-silver contraption he calls The Dominator as it works its way across a warehouse floor. Built for one purpose—setting up dominoes—it neatly laid out more than 100,000 tiles in under 24 hours, obliterating the human world record.[7]

The Dominator is a marvel of specialization on a niche task, yes—but only in certain conditions: perfectly flat floor, no stray Lego bricks, zero gusts of air, and a particular domino size. Add a carpet seam or a curious cat and that purpose-built speedster turns into an expensive paperweight. The Dominator is a reminder that raw efficiency collapses the moment the environment wanders off-script.

While engineering flexibility into a domino-placing robot might be technically feasible, the price tag quickly spirals out of control. Pouring resources into such adaptability for a highly specialized task simply doesn't make economic sense.

Now pivot to AI, which aspires to be both conductor and jazz soloist—precise yet spontaneous—but still stumbles on the latter. Early self-driving car prototypes cruised confidently down sunny California highways yet froze when reality served up something un-American—for example, a kangaroo when being tested in Australia. The bounding marsupial broke the object-detection math because it hopped above the road, then landed closer than expected, baffling sensors trained on moose, elk, and pedestrians.[8]

Fixing quirks like the kangaroo problem isn't cheap. By 2020, companies had invested roughly $16 billion in robo-taxis, with Alphabet's (Google's) Waymo alone burning through billions before its first paid ride.[9] That budget makes sense when the prize is a trillion-dollar transportation market. But most industries don't dangle jackpots big enough to justify that amount of investment—not for a domino-laying robot, anyway.

What is the takeaway? Specialize like a machine but adapt like a mammal. Your narrow expertise keeps algorithmic competitors out—they won't spend $16 billion to address the edge cases for unusual tasks when your animal flexibility handles the curveballs far less expensively. When the market is too tiny for OpenAI to care about but too messy for a simple algorithm to thrive—think kangaroos—then you become irreplaceable by default. That's the Kangaroo Defense.

Consider specialty electricians. Rope-access electricians dangle from rock-climbing gear to fix lighting and wiring in stadium rafters, never knowing what surprises lurk behind the walls. Industrial-ride electricians keep roller coasters and theme-park machinery alive, troubleshooting one-of-a-kind systems. These roles are so niche, so physically particular, and so poorly documented that it's hard to imagine AI gathering enough training data to replace them.

Or go even weirder: forensic locksmiths, underwater welders, horologists who revive antique watches, air-accident investigators piecing together clues from mangled aluminum. If AI transforms or eliminates many common jobs, the world may end up populated with far more of these ultra-specialized, curveball-filled roles, the kinds of odd careers that stay safely out of reach precisely because they're too small, too strange, and too unpredictable to automate.

Still Worried Your Career Won't Adapt? Relax, There's a Solution

If this chapter and the previous chapter left you eyeing your paycheck a little nervously, take heart: The next one is less about (possible) career doom and more about how to respond if your career is targeted by AI. The answer lies in one word: reskilling. Reskilling doesn't just pad your résumé and wallet; it can also protect your well-being. In one study of workers sidelined by old-fashioned, non-AI injuries, retraining cut the risk of depression sharply. Specifically, for every three workers reskilled, one case of depression was prevented. Overall, access to reskilling cut the rate of new antidepressant use by about 70%.[10]

Fortunately, reskilling is becoming easier than ever. Silicon competitors may replace some of today's jobs, but they're also turbocharging the very thing that makes tomorrow's jobs possible: retraining. Your next career could be closer than you think—perhaps just one chapter away. Read on to learn how.

4
The Education Rip-Off
Will AI Finally Pop the Tuition Bubble?

Here's a question that is too harsh to ask at back-to-school night: What if all the extra money we've poured into education hasn't actually made us any smarter than our parents? Before we can explain how technology might finally transform education and reskilling, we first need to understand why education was structurally destined to fail long before AI showed up. To do that, we're taking a field trip backward through time. Way back. Further than you might expect.

Why Ordinary Schooling Keeps Getting Pricier: From Toga Times-Tables to Six-Figure Tuition

Picture yourself in ancient Rome. You're perched on a backless wooden stool, wax tablet balanced awkwardly on your knees, while a weary schoolmaster loudly drills you on grammar at dawn in a rented storefront called a *pergula*. Classes mix kids of every age, the teacher hops from one pupil to the next, and parents pay tuition in awkward installments, when they remember to pay at all.[1]

Fast-forward two millennia and swap the toga for a Tommy Hilfiger hoodie. The furniture has cushions now, but the instructional model is eerily unchanged: One adult voices facts to a semicircle of fidgety learners, each moving at a different speed. This surprising continuity in educational technique explains why schooling feels stuck in the past and why the price tag keeps climbing.

BAUMOL'S AWKWARD MATH LESSON

Economist William Baumol spotted this problem half a century ago, and it still defines education today. In most sectors, workers produce more over time: more vehicles per autoworker, more lines of code per programmer, more garments per seamstress. Rising productivity often supports rising wages.

Education defies this pattern. In 1955, a U.S. teacher managed roughly thirty students. Today, that number has dropped to about fifteen.[2] Smaller class sizes may indeed produce better outcomes. But by the standard metric of output per worker, each teacher now "produces" half as many graduates. Productivity does not appear to be rising. If measured per student, it instead is falling.

Yet teachers exist in the same labor market as everyone else. Schools must offer salaries competitive enough to attract talent away from nursing, accounting, or engineering. Teachers may accept somewhat lower pay to pursue their calling, but there's a limit. Without competitive compensation, schools face shortages and wages rise.

The problem? Productivity doesn't increase alongside wages. This imbalance is what Baumol termed the Cost Disease. Schools aren't wasteful or inefficient; they're simply trapped in work that resists efficiency gains. As wages rise across the economy, education costs climb, but without productivity improvements to absorb the added expense. This structure virtually guarantees that cost of education will keep rising.

Sticker Shock, 2025 Edition

In fall 2025, Wellesley College invoiced families $100,541 for a single year of tuition, room, board, and the rest of campus life's trimmings. Wellesley isn't an outlier; the annual costs of attending a pack of private colleges, including Tufts University and Vanderbilt University, now hovers near the six-figure line.[3] K-12 districts feel the same squeeze: Salaries rise to keep teachers from decamping to higher-pay sectors, while class sizes remain stubbornly small by law, union contract, or choice.

In just ten years, between 2011 and 2021, K-12 spending per pupil rose by about a third.[4] Even after accounting for inflation—the fact that prices in general go up over time—schools are still spending about 16% more per student than they did a decade ago. It's one of education's most frustrating contradictions—we keep spending more without getting better results.

The Elephant in the Classroom

The price of education isn't the only problem. Pace is another obstacle. Any given lesson is Goldilocks-sized for maybe a third of the room: too slow for the quick, too fast for the struggler. Districts pour cash into enrichment for the bored and remediation for the bewildered, but the factory-model classroom retains this manufacturing mismatch.

So the next time you gasp at a $100,000 tuition bill or the amount you pay in local school taxes, remember the through-line:

We're paying 21st-century wages for a 3rd-century BCE production process. Unless we reinvent learning one child at a time, the educational system will keep alchemizing ordinary chalk into 24-karat tuition.

PROOF OF CONCEPT: FLESH-AND-BLOOD TUTORS WORK MIRACLES—AND THAT'S A PROBLEM: BLOOM'S TWO-SIGMA BOMBSHELL

Benjamin Bloom wasn't a typical education researcher. After earning his Ph.D. at the University of Chicago, he became the university's official examiner—a position few professors today would envy. His job was to make sure that students' final exams actually demonstrated mastery of what they were supposed to have learned. It was an unusual post, but it gave Bloom a front-row seat to observe one of education's enduring mysteries: what real learning entails, and why some students thrive while others fall behind.

Bloom's studies led to a discovery that would become one of the most famous findings in modern education. He showed that pairing a student one-on-one with a skilled tutor boosted the student's test scores dramatically—by two standard deviations above what students typically scored in a regular classroom.[5] One standard deviation (known in statistics as "sigma" and represented by the Greek letter σ) represents a large leap in performance, so two full sigmas translate to an enormous effect: enough to catapult an average, middle-of-the-pack student straight onto the honor roll. In defining the *two-sigma problem*, Bloom highlighted not only the huge size of this tutoring effect but also the frustrating question it immediately raised: How can we possibly afford to give every child on Earth their own personal tutor?

The Modern Scoreboard

Four decades and 96 randomized trials after Bloom published his results, the scoreboard still favors tutors. A 2020 NBER meta-analysis by Andre Nickow and colleagues finds that one-to-one or tight-knit small-group tutoring delivers an average 0.37 sigma bump—basically a year of extra learning crammed into a semester.[6] In addition, dosage determines destiny: Students tutored daily or three times weekly during school hours see nearly double the gains of those who get once-a-week or after-school help. Teacher-led sessions beat parent-led sessions. The fiscally painful conclusion? The most effective tutoring models are also the most expensive.

Why the Gains Don't Scale

Although the evidence is overwhelming, it's clear that the United States is not drowning in tutors. Two frictions keep Bloom's two-sigma miracle from being realized:

1. *Unit labor costs.* Even paraprofessionals earn non-trivial wages. Districts that pay $70,000 for a veteran teacher balk at tacking on another $35,000–$50,000 of tutoring staff per classroom. Policymakers view tutoring as too costly to undertake on a large scale.
2. *Human supply chains.* Saga Education (a tutoring service) can hire a thousand recent grads, but the minute you need 10,000 algebra tutors, the recruiting funnel chokes.

The upshot? Human tutoring is a proven but rationed luxury. It is a vitamin we know works, yet dispense with an eyedropper because the price of mass production is prohibitive.

Benin City, Nigeria: When a Chatbot Cracked Open the Classroom Door

Picture a sweltering afternoon in Benin City in Nigeria. Forty teenagers share eight battered desks; the lone English teacher is juggling flood warnings and a curriculum that gallops faster than half her pupils can read. Then comes an announcement straight out of science fiction: *School just got a 90-minute after-hours add-on, and students now have a new study buddy that is a chatbot built on Microsoft's flavor of ChatGPT.* "AI helps us to learn, it can serve as a tutor, it can be anything you want it to be, depending on the prompt you write," says Uyi, a student at Edo Boys High School, as he logs in for a session.[7]

Six weeks later, the numbers looked like a rigged video-game score. The 250 kids who chatted with the bot outscored their control-group classmates in their English studies by 0.30 standard deviations—roughly two full years of normal learning crammed into 42 days. Even better, the bonus spilled over into end-of-year exams in subjects the bot never "taught."[8] Attendance worked like compound interest. Each extra day students participated in the program nudged scores higher, and the gains never plateaued.

Now consider Rising Academies, a nonprofit organization founded in Western Africa in 2014 to create the world's best schools and to service underserved communities. Partnering with Anthropic's Claude, Rising Academies rolled out Rori, a WhatsApp math tutor, and Tari, an AI coach for teachers. Together they already reach 150,000 learners across five countries and clock a 0.30 σ jump in math mastery—at data costs low enough to run on dollar-a-gigabyte cell plans.[9]

What ties these stories together isn't just flashy effect sizes; it's the way a handful of code lines can morph into 24/7, polyglot, infinitely patient tutors. In a system afflicted by Baumol's Cost Disease, that's nothing short of educational counter-alchemy, turning overcrowded chalk-dust misery into gold-standard learning, one prompt at a time.

But—Pump the brakes. *The Economist* magazine warned that such moonshots are hardest to repeat in rich-world classrooms already flush with teachers and support staff.[10] In other words, AI may yield its biggest dividends where the human tutor shortage is most acute. It isn't (yet) a viable strategy for the developed world.

Still, technology companies are sniffing around. AllDayTA, which digests a professor's slides and handouts and then moonlights as a virtual TA (teaching assistant), is an early experiment examining whether AI can substitute for teaching assistants or professors in providing certain types of support. OpenAI's "study mode" could nudge things further by utilizing ChatGPT's cutting-edge models. And then there's Alpha, a private school claiming that its AI-plus-app combo lets kids cover the curriculum twice as fast, reducing classroom instruction to two hours a day and freeing time for group work and extracurricular activities.[11] If Alpha's claims are true, then we're not looking at a school, we're looking at a prototype for reinvention.

There's a catch, though: Just like with Zoom school, the returns may split along motivation lines. Engaged students rocket ahead; disengaged ones drift further away. The real frontier isn't the chatbot's IQ, but whether it can keep even the slackers from logging off. Keeping unmotivated students engaged could make AI tutoring more than just a fix for the developing world—it could make it a global education game-changer.

Can Education Become Addictive? When Dopamine Hits Serve Learning Instead of Scrolling

The same month Uyi was chatting with his English tutor in Benin City, American teenagers were averaging 4.8 hours daily on social media—not for homework, but for scrolling through TikTok, Instagram, and YouTube.[12] The platforms deploying the most sophisticated behavioral psychology on Earth aren't trying to educate; they are harvesting attention like a cash crop.

Silicon Valley's addiction algorithms are masterclasses in human motivation. Every swipe, pause, and double-tap feeds machine learning models that map your neural reward pathways with surgical precision. The techniques are well-documented: Variable reward schedules mirror slot machines—you never know if the next video will be pure gold or merely good, so you keep pulling the lever. Social-validation loops turn likes and comments into digital cocaine hits. Fear of missing out (FOMO) keeps notifications buzzing at bedtime. Infinite scroll eliminates natural stopping points, like a casino floor with no clocks or windows.

What if we flipped the script? Today's social media platforms use "dopamine hacking"—tactics that tap into the brain's reward system to keep us scrolling one more video or checking one more notification. But what if those same algorithms were optimized not just to hold attention, but to boost test performance and sustain a student's focus? Imagine an AI tutor whose goal isn't only to answer your calculus question, but also to make you crave the next problem set.

Some organizations are already experimenting. For example, Khan Academy uses goals, progress bars, and achievement badges to enhance student motivation.[13] But these tools are arguably kindergarten-level compared to what Meta (parent company of Facebook), YouTube, and ByteDance (TikTok's owner) subtly deploy via their proprietary algorithms.

The Test-Score Feedback Loop

Education has an objective benchmark: actual learning outcomes. An AI tutor trained to maximize a student's performance on a standardized test six months later would encounter the fascinating challenge of optimizing the knowledge transferred to the student. This challenge would require the algorithm not only to optimize how it delivers material but also how it sustains student attention; both are central to effective learning. With each additional user, the system would accumulate evidence, iteratively improving its ability to teach.

What are some possible outcomes? The system might discover that occasional moderate spikes in difficulty sustain long-term retention better than constant validation from easy problems. It might learn that spaced repetition, though less immediately gratifying than new content, pays dividends on the final exam. In other words, the algorithm could learn to become both a better teacher and a more compelling attention hog.

The potential upside is staggering. If we could redirect even a fraction of Americans' daily screen time from consumption to learning—if we could make calculus as compulsive as cat videos—we might solve Bloom's two-sigma problem not only by imitating human tutors, but also by applying the same behavioral psychology that currently keeps us scrolling past midnight.

That said, my colleagues vehemently believe that students focus better with human teachers, or something closely resembling them, because there's something about a face and a voice—even synthetic faces and voices—that captures attention in ways pure text cannot. So, if we're going to use the tools of behavioral psychology for education, why not give our AI tutors a human form?

"Lights, Camera, Algorithm!"—Hollywood in a Browser

You might ask: If students respond better to human faces, why use AI at all? The answer is straightforward—humans are expensive and imperfect. The ideal classroom of the future will combine the best of both worlds: human warmth and judgment paired with machine precision and adaptability.

My own struggles prove the point. Last night I rehearsed tomorrow's in-person econ lecture. The script on my laptop crackled with wit; the words dropping from my mouth, not so much. To optimize my students' learning experience, I might create a video by recording a dozen takes and splicing the best bits together. Or I could use a service that some companies now offer, pasting the script into an AI video generator and letting a silicon doppelgänger nail it in one take. This approach sacrifices the benefits of in-person lecturing but compensates with flawlessly chosen wording.

Tools such as Synthesia and HeyGen turn plain text into a full-HD talking head, complete with gestures, slides, and studio lighting. No microphone, no green-screen, and no awkward reshoots are required. The user interface feels like PowerPoint with a "make me talk" button: Write the lecture, pick an avatar (your clone or one of the stock presenters available in the software), hit render, and sip coffee while the cloud sweats over the pixels.

- Time cost: A five-minute video takes about as long to generate as it takes you to read the script aloud—once.
- Cash cost: Many platforms charge $1–$3 per finished minute. A semester's worth of 12-minute micro-lectures costs less than a single hour in a campus studio.
- Revisions and updated content: Edits are nearly effortless. If a slide changes, swap the text and click *regenerate*—your avatar never complains.

Deepfake Professors, Real Pedagogy

Now imagine that your avatar doesn't just parrot your baseline lecture but also *personalizes* it. Large language models already condense, expand, or translate your notes on command. Combine that with a video generator and you get a tutor that:

1. Speaks 140 languages on demand. One click and your macro-econ example about Zimbabwean hyperinflation is delivered in Shona, the most widely used language in Zimbabwe.
2. Paces to each viewer. Struggling with supply-and-demand graphs? The bot inserts extra scaffold slides. Racing ahead? It unlocks an optional digression on an advanced topic.
3. Remembers misconceptions. Missed the minus sign in last week's elasticity quiz? Expect a bespoke micro-review stitched into tomorrow's clip.

Marginal cost of Version #2 for Student #2,000? Practically $0.

The Tutoring Frontier, Upgraded

As we've seen, three main features make the future of education exciting.

- *Marginal cost ≈ $0.* Once the engine is up and running, serving Lagos, Louisville, and Lahore simultaneously costs pocket change.
- *Stronger student engagement.* Algorithms designed to maximize student performance improve both pedagogy and entertainment value to keep students engaged.

- *Infinite differentiation.* One master script can spawn 8th-grade, AP, and ESL versions of the same content, each tagged to a different learning profile.

In other words, AI video won't just put your lecture on autopilot. Instead, it will harness Bloom's insight that *individual pacing trumps industrial pacing*, but without adding $4,000 per semester to the tuition bill.

And it's about time. For roughly 2,000 years—from Aristotle's Lyceum to your kid's chemistry class—the basic model hasn't budged. One teacher, many students, everyone moving at the same glacial pace. The printing press came and went. The internet arrived and mostly just digitized the same old lecture halls. But now, after two millennia of educational stagnation, we may be staring down the barrel of the first genuine educational breakthrough since someone figured out how to write cuneiform on a clay tablet.

BEYOND THE CLASSROOM: WHEN AI HITS THE STREETS

Just as AI is gearing up to shatter the ancient classroom mold, another technological revolution is quietly humming at the curbside—one that promises smoother rides but uncertain destinations. What happens when technology such as self-driving cars doesn't just alter your commute but also reshapes entire cities, redraws economic fault lines, and sets off political shockwaves? In the next chapter, we examine how your driverless future might have more potholes (and gunshots) than advertised.

5
Total Inequality
Is the Future Utopia for the Rich, Dystopia for the Rest?

From Cloud to Curb Stomp

Earlier, we tackled the obvious questions: "Will AI take my job?" and "How can I reposition myself?" This chapter goes deeper—into the ripple effects. Even if you're completely immune to direct job loss, the indirect impacts will reach you. And some of them are far more unsettling than the direct effects. We'll begin with what's aggravating, confront what's darker, and then explore how we might respond.

The Man Who Drove the Future

Many people believe that self-driving cars were born in the 2010s, when Silicon Valley engineers and venture capitalists decided to teach cars to think. But the real story starts long before Tesla's

Autopilot or Google's Waymo, with a German aerospace engineer named Ernst Dickmanns, who quietly built the first truly autonomous vehicle decades earlier.

Back in the 1970s, Dickmanns had worked with NASA on automated landing systems for airplanes.[1] The experience left him with a radical idea: If a jet could guide itself onto a runway, why couldn't a car guide itself down a highway? Most dismissed the notion as absurd. Airplanes operate in open skies under controlled conditions; cars face pedestrians, traffic lights, and potholes. "Autonomous driving" sounded more like science fiction than science.

But, with the encouragement and funding of Mercedes-Benz, Dickmanns persisted. By 1994, he and his research team had wired a laptop into a modified van and trained it to recognize lane markings and steer accordingly. On a bright day on the autobahn, it did exactly that, cruising at 60 miles per hour, completely hands-free. Cameras watched the road. Software made the decisions. For a moment, the future had arrived.

There was, however, one catch: The autobahn was empty. Later trials on public roads achieved partial success, but the risks were enormous, and computing power at that time was woefully limited. Still, the technology didn't vanish. Instead, it evolved sideways. The algorithms that once steered Dickmanns' five-ton van began showing up in more practical features: lane-departure warnings, adaptive cruise control, and collision-avoidance systems. These safer, narrower applications quietly entered vehicles while the dream of full autonomy was set aside. The world wasn't yet ready to hand over the steering wheel to a machine.

Two decades later, the world caught up. Faster chips, high-resolution cameras, and LiDAR sensors capable of mapping the world at centimeter-level precision suddenly made Dickmanns' dream feasible. What once required a room-sized computer, machine learning could now do in milliseconds. By the 2010s,

Tesla's Autopilot was navigating real traffic, and Waymo's robotaxis were chauffeuring riders across Phoenix. We had gone from "That's impossible" to "Can I nap during rush hour?" in a single generation.

The implications extend far beyond convenience and the five million driving jobs that economists expect to disappear in coming years. When driving no longer requires our attention, the nature of mobility itself changes. Parents may send their children safely across town without acting as unpaid chauffeurs. The elderly and visually impaired could regain freedom of movement. Perhaps most profoundly, millions of commuters may no longer be bound by the geography of traffic. Where we live, where we work, and how cities are designed all become newly negotiable.

Dickmanns couldn't have known it then, but that quiet autobahn experiment in 1994 wasn't just a technological milestone. It was the opening move in a slow, seismic reordering of human activity. When we no longer have to focus on the road ahead, we start reimagining where that road can lead.

Imke Reimers and I forecast the impacts in a recent paper.[2] Because the robo-chauffeur hasn't fully taken off yet, we reverse-engineered its appeal. First, we measured what workers hate about today's commute: white-knuckle steering, missed transfers, train delays. Then we asked, *What if every ride felt like door-to-door first-class rail, with no transfers, no delays, and no merging mayhem?* Basically, we imagined your own personal Uber driver already waiting in the driveway, minus the small talk and the surge pricing, at about the same cost as owning a car today. The goal was to duplicate public transit's ace in the hole—the commuter's ability to read a novel while somebody else drives.

Suddenly the world is full of new possibilities. Why live in a shoebox near the office when you can snooze through a 90-minute commute? Punch in your destination, crawl under a blanket, and let your personal robo-driver convey you to the office at 65 mph.

Our empirical model predicts that the average urban worker will bolt for the 'burbs, crank their daily mileage up 40%, and drain 10 to 13% of public transit's ticket revenue, even after accounting for slowdowns from the resulting increased road congestion. The typical home-to-office distance stretches (see image below); downtown speeds decrease and so do speeds in nearby suburbs. Conclusion: The moment your car learns to drive itself, you start treating the highway like Hulu—something to binge.

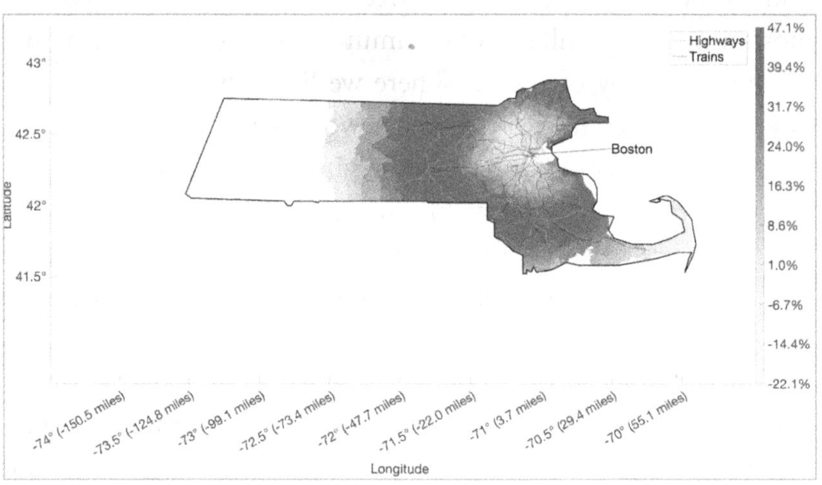

Figure 2. *How Self-Driving Cars Could Shift Where People Live*

Figure Notes: This map uses color hues to depict projected changes in population resulting from the advent of self-driving vehicles. Blue areas represent population increases, while yellow tones indicate declines. Light gray areas show little or no change. Overall, the map suggests that about 10%–20% of Bostonians are expected to move out of the city, with population growth of roughly 25%–50% concentrated in suburbs about 15 to 25 miles away.

And that's before the delivery bots and joyrides join the party. Translation: Your car might be hands-free, but your highway will be elbow-to-elbow. Without smart fixes—such as congestion pricing, bus-only lanes, maybe a few fresh miles of asphalt—the traffic apocalypse likely spreads far beyond Manhattan and other major metropolitan areas.

Earlier studies reveal a familiar paradox: When cities expand their highways to ease congestion, the relief is short-lived.[3] Housing patterns shift, people move farther out, and commute times return to where they started. With self-driving cars, a similar dynamic seems likely, but instead of keeping drive times the same, it will keep the associated hassle unchanged. As the drudgery of driving fades, consumers may simply stretch their commutes, trading shorter drives for longer, easier ones. The hassle per minute drops, but the total time spent behind the wheel might rise, leaving us no better off than before. Progress, meet your unintended consequence.

So did all that innovation just bring us full circle? Hardly. We haven't yet looked at the ripple effects, where the real story begins.

THE DOMINO EFFECT: FROM ASPHALT TO INEQUALITY

Zoom out from the morning commute and you'll notice that AI won't just devour lane-miles. It will also take a bite out of paychecks, and possibly (metaphorically) load a gun.

Let's start with wages. MIT's Daron Acemoglu and Boston University's Pascual Restrepo sifted through forty years of U.S. payroll data and reached a startling conclusion: Automation alone explains roughly 50% to 70% of the increase in wage inequality since 1980. For men without a high school diploma, the tasks in which they once had a comparative advantage were disproportionately automated, diminishing their job prospects and leaving their real earnings about 15% lower than what their fathers earned at the dawn of MTV.[4] AI could accelerate this trend. Erik Brynjolfsson paints a bleak picture of what he calls the *Turing Trap*: For some workers, AI doesn't just compete—it dominates. When machines can do everything a person can do, but faster, at lower cost, and without complaints or sick days, labor prospects evaporate.[5] Self-driving vehicles offer just one vivid example. They promise safer roads but threaten to sideline millions of truckers, taxi drivers, and delivery workers whose livelihoods depend on steering wheels. Even if certain jobs remain human-dependent, the surge in displaced workers vying for them will swell labor supply and push wages down.[6]

A thinner wallet doesn't just wreck the dinner budget. It also warps the social fabric. A recent NBER paper titled *Robots and Crime* linked local levels of robotic automation to corresponding increases in local crime rates. The bottom line: A 10% increase in local robot adoption is associated with a 0.2%–0.3% rise in property crime. Nationwide, robotic automation has added an estimated $322 million in additional crime-related costs.[7]

History suggests that violence may also increase with automation. In a classic cross-country study, economists Pablo Fajnzylber, Daniel Lederman, and Norman Loayza showed that when the gap between rich and poor people gets wider, violence tends to rise, even after accounting for poverty, policing, and a dozen other confounding variables.[8] Inequality, it turns out, is a remarkably efficient crime accelerant. And the wider the gap, the shorter the fuse.

Put bluntly: An algorithm isn't typically pulling the trigger, but the pink slip it prints can load the chamber.

How Might the Public Push Back?

June 2025 gave us a preview of the worker pushback—and panic—that AI would stir among creators. Seventy prominent authors, from Lauren Groff to R.F. Kuang, published an open letter demanding that publishers swear off AI-generated prose, voices, and cover art.[9] Their gripe isn't literary snobbery; it's the fear of becoming "training data with a mortgage"—that is, humans whose life's work is scraped by AI companies to train the very AI models that then undercut their livelihood. If award-winning authors feel threatened, imagine assembly-line workers staring at a Boston Dynamics robotics arm that never calls in sick.

This book could be Exhibit A in the prosecution: a draft hammered out in weeks thanks to large language models, with images generated by AI. Multiply that productivity jolt across history wonks, Lego-set hobbyists, or someone dying to write *Macroeconomics of Martian Muesli*, and the Kindle store will flood with titles crowding out traditional authors. Few of these books may ever be read, but when the time cost of creation drops, output explodes. Many authors will be writing mainly for friends or family—but every upload still crowds the same digital shelf.* This is great for readers but unnerving for writers whose advance just evaporated.

Will open letters and public outrage stop AI from reshaping authorship? Possibly, but don't bet your book advance on it.

Plan B for anxious workers is the ballot box. The results of many U.S. elections are razor-thin. For example, Georgia flipped in 2020 by 0.24 percentage points, Michigan by 0.63 percentage points, and eight other states have seen margins under 1% since 2000.[10] When the Electoral College can hinge on ~80,000 disgruntled voters, even a modest cohort of automation casualties could swing the presidency—and thus policy—toward slowing or shackling AI.

---

* I have found it far easier to write fiction than nonfiction. For example, I use AI to write books starring my kids as the heroes of the stories. Each 50-ish page book took only 30 minutes to write. They are not as good as human-authored children's books, and they might not have always matched my intent and tone, but they more than made up for it in their personalization. My kids read them over and over.

Or they might reach for a nineteenth-century playbook. The original Luddites (1811–1816) smashed machinery they feared would starve their families; Parliament answered with mass arrests and, in one case, executions.[11] No one's forecasting pitchforks at Nvidia headquarters just yet, but let's not rule out the 21st-century equivalents: cyberattacks, flaming Waymos, protest blockades outside data centers, or thousands of protestors intentionally blocking traffic.

Overall, none of these solutions is particularly satisfying. They may preserve jobs, but only by slowing progress.

CORPORATE PREEMPTION: KEEPING THE CROWN JEWELS IN HOUSE

Firms can read the room, too. OpenAI's board has already given its new Safety Oversight Committee the power to delay or block new models if they pose serious risks. The simplest risk-reducing move is to stop shipping your best models and use them privately, depriving startups of the same uplift. That strategy might shrink the total welfare pie, and maybe even OpenAI's long-run revenues, but it buys political insurance against the backlash described above. But just like the workers who take defensive action, firms keeping AI development in-house to avoid public backlash are doing so at the expense of progress.

HOW CAN WE REAP THE GAINS AND DODGE THE PAIN?

The brutal truth is that automation creates winners and losers. The winners buy islands. The losers buy guns. How do we manage this unsustainable equilibrium?

Option 1: Tough Luck, Deal with It

While I disagree, some say the solution is "Do nothing." They argue that disruption works. Would we be better off if 80% of us still pushed plows and stored food for the winter? Displaced farmers in 1850 lost their livelihoods, but their misery bought us antibiotics, air conditioning, and smartphones. In other words, we traded subsistence farming for a world in which people in developed countries often spend more energy trying to shed pounds than worrying about where their next meal will come from. Where their greatest entertainment dilemma is choosing among 500 Netflix shows. Where they complain about Wi-Fi speed while flying through the sky in metal tubes. Every one of these advances required someone, somewhere, to lose their job to a machine.

Like evolution, the market is brutal but efficient. Farmers became factory workers, who became service workers, who will become ... whatever comes next. Yes, individuals get crushed in the gears of the economy. Yes, entire towns hollow out. Yes, we might need to beef up the police force when the taxi drivers realize their costly medallions are worthless paper.*

But the hand-wringers miss an important insight: The alternative leans more toward stagnation. Every job we protect is an innovation we delay. Every worker we shield from disruption is a breakthrough that doesn't happen. The choice isn't between disruption and stability—it's between painful progress and comfortable decline.

So maybe the answer is exactly what it's always been: Let creative destruction create and destroy. The destroyed rarely thank you, but their grandchildren will. However, I'm one of many who

* In New York City, taxi medallions—which once sold for over a million dollars each—lost about 80% of their value after Uber arrived. Self-driving cars could wipe out what little value remains in a taxi medallion.

believe that easing the transition for those swept up in this tidal wave is crucial. Even better, this solution won't slow down growth much.

Option 2: Universal Basic Income

The second option is to issue everyone a monthly check, something like what Andrew Yang proposed during his 2020 presidential campaign. Yes, even Elon Musk gets one. The amount? Enough to keep you housed and fed, but not enough to buy a Tesla—an amount low enough that the total cost could easily be covered by taxes on the productivity gains unleashed by AI. Limited trials suggest that universal basic income (UBI) can be a good strategy, challenging the idea that free money creates free-loaders. Those receiving the check often continue to work.[12] Humans actually like having purpose.

Option 3: Progressive Taxation and Redistribution

The third option is Robin Hood, but with spreadsheets instead of arrows. Instead of sending a check to everyone, as in the case of universal basic income, send a monthly check only to the less fortunate. For some, progressive taxation and redistribution seem fairer. In truth, though, Options 2 and 3 are basically the same thing. Once you factor in who is paying for the universal income —the haves—the logic crystallizes.

The transformation from progressive taxation to UBI is straightforward: Increase taxes to fund the checks to the haves as well as the have-nots, then distribute universal payments. For wealthy individuals, this system would create a circular flow—they pay larger sums to the government, then receive small amounts back. If the government wants to send Elon Musk a $20,000 check, it can simply raise his taxes by $20,000 to fund it. His net position remains unchanged.

Option 4: Guaranteed Jobs

The fourth option, supported by Bernie Sanders, is a federal job guarantee: offering employment to anyone who needs it.[13] The intent is not to replace private-sector work, but to provide a safety net for those unable to find jobs elsewhere, so long as they are willing to put in the work hours.

Options 2 through 4 share one key similarity: They involve taxing the rich more in order to ensure the less fortunate can afford basic necessities. But won't the rich revolt? Sure, they'll whine on X, but they won't revolt. Instead, they should be writing thank-you notes to Bernie Sanders and Andrew Yang for promoting income redistribution. Why? Because angry mobs are bad for property values. Gated communities have lousy ROIs if the rest of the city burns. And a CEO shot dead on the sidewalk enjoys exactly zero yachts. That example isn't made up. UnitedHealth's CEO, Brian Thompson, was killed in public in 2024—a stark reminder of how resentment over inequality can turn violent.[14] A functional society isn't socialism. It's a valuable insurance policy.

But wait. Do we need to tax the rich? What if we're thinking about this backwards? Why don't we just tax the bots? If a robot takes your job, shouldn't the robot pay your unemployment benefit?

Should We Just "Tax the Bots?"

On the face of it, funding a nationwide universal basic income by slapping levies on robots and large language models has a pleasing symmetry: If automation kills the job, then automation pays the rent. Bill Gates pitched exactly that logic, arguing that we already tax human labor, so why give silicon a free ride? That's a nice sound bite but messy economics. Why?

- *Every tax distorts behavior.* Put a tax on GPUs or per-algorithm transactions and firms will use fewer of them. Such taxes slow the progress of the technology poised to fatten future GDP. We should be subsidizing AI technology, not taxing it.
- *Capital chases low-tax regions.* If the United States imposes a "bot toll" while, say, Shenzhen, China, rolls out a red-carpet rebate, venture dollars and engineering talent will follow the less expensive runway. The result will be fewer home-grown AI giants, more foreign ones, and a smaller tax base to tap later.
- *Broad taxes work better than narrow taxes.* Classic public-finance math says the least harmful way to raise a dollar is with a broad, low-rate tax (such as a value-added tax) or Pigouvian taxes (such as a modest carbon tax), not a bull's-eye on a single, fast-evolving sector.

Here's the bottom line: Universal basic income may be the right parachute, but taxing only the bots is a flimsy ripcord whose main virtue is soothing public anxiety. If we want a social safety net without kneecapping innovation—or handing the next industrial revolution to our geopolitical rivals—we need a wider revenue lens than "Make Siri pay payroll tax."

Choose Wisely or Pay Dearly

Dickmanns' 1994 self-driving Mercedes was a simple proof of concept: Silicon could match human reflexes at 65 mph. No harm done, just a computer keeping a car between the lines. But thirty years later, that same technology isn't just steering rideshare vehicles in San Francisco, Phoenix, and beyond—it's steering our entire economy. The stakes have changed. When AI controls both your commute and your career prospects, getting the rules wrong means more than a fender bender. It means societal wreckage.

The possibility of this wreckage leads us to the economist's prime directive: Craft rules that hit the target without letting the shrapnel wound everything else. History overflows with noble laws that boomeranged on the very people they meant to help. The classic example—seen vividly in Venezuela—is a government capping food prices to keep them affordable, then printing money to cover losses. When inflation pushed supplier costs higher while a price ceiling kept prices low, losses became inevitable. Many suppliers shut down rather than operate at a guaranteed loss. The result was empty shelves and hunger. Taxing the bots may be similarly destructive. The lesson: We need politicians focused on policies that actually work, even if they're not the easiest to sell at rallies.

Playing with Fire: How Close Is Too Close?

Self-driving cars, economic upheaval, and suburban sprawl—these changes seem world-altering until you realize they might just be warm-ups. In the next chapter we'll discover whether the greatest risk is that AI will give us exactly what we accidentally ask for, thereby keeping the universe eerily quiet. Confused by that last statement? Don't worry—by the end of the next chapter, it'll all make sense, even if it terrifies you.

6
Is AI Going to Kill Us All (Eventually)?
Disturbingly Sound Logic

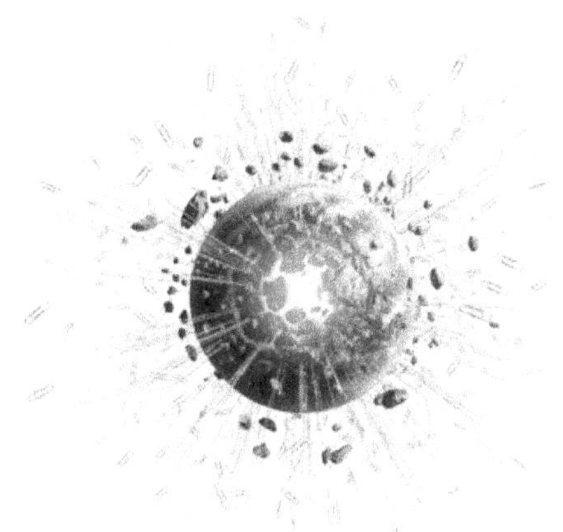

This chapter examines logic suggesting our impending demise that's uncomfortably sound—indeed, sound enough to keep me awake at night. And yet, while easy solutions exist, I'm not sure we should use them. To understand why, we need to weave through a few different ideas, starting with stories about aliens (the outer-space kind). And before you accuse this chapter of jumping the shark—that is, veering suddenly into absurd territory—trust me, it all connects.

Cosmic Silence and the Red Eye Encounter

Imagine this: You're stuck on a red-eye flight next to someone convinced he's been abducted by aliens, probed, and handed cosmic secrets. You crave sleep, but curiosity whispers, "What if he's right?" Even the most rational economist can't completely ignore such intriguing slivers of uncertainty.

That nagging "What if aliens exist?" question isn't unique to sleep-deprived flights. Pop culture loves alien theories, from Area 51 memes to Men in Black gags. Yet, after decades of diligent searching, SETI (the Search for Extraterrestrial Intelligence) hasn't picked up even a whisper from space. Complete silence. Does this mean humans are utterly alone in the universe? It sounds plausible, but then again—does it really?

You're probably asking yourself what any of this has to do with AI. The answer is everything, though not in the way you'd imagine. To understand the connection we need to take a detour through the cosmos.

The Fermi Paradox and Cosmic Mysteries

On a sweltering Los Alamos lunch break in 1950, Nobel laureate Enrico Fermi dropped his fork and lobbed a cosmic grenade: "So ... where are they?"[1] He wasn't talking about tardy colleagues; he meant alien civilizations. Let's do the math: roughly 100 billion galaxies, each boasting billions of stars—that's on the order of 10^{22} suns with potential real estate for life. How big is 10^{22}? That number is roughly equal to the total number of grains of sand on every beach and desert on Earth, multiplied thousands of times over. Given the staggering number of stars and planets, shouldn't at least *somebody* be texting us by now?

A decade later, radio astronomer Frank Drake transformed Fermi's lunch-table musing into something more systematic. His now-famous equation chains together seven cosmic numbers and probabilities, each one a necessary hurdle for intelligent life to arise elsewhere and make contact with us. Start with the rate of star formation in our galaxy. Multiply by the fraction of those stars with planets. Then multiply by the average number of potentially habitable planets per star system—the planets in the "Goldilocks zone" where water can stay liquid. Keep multiplying: by the chance life actually sparks into existence, by the chance it evolves intelligence, by the probability that intelligence develops broadcasting technology, and finally by how long such civilizations keep their transmitters running before going silent.

Note that these aren't additions; they're multiplications. And here's the brutal mathematics of multiplication: If any single factor in the chain is exceedingly rare—close to zero—it poisons the entire equation. It doesn't matter if every other probability is high. You could have a billion planets in the habitable zone, but if the jump from chemistry to life happens only once in a trillion attempts, the final number of planets with intelligent life still collapses to nearly zero. One near-impossible step drags everything down with it, rendering all the other promising factors irrelevant.

So what does the math tell us? If intelligent civilizations were abundant, then the cosmos should be buzzing with signals. Yet we hear nothing: no transmissions, no megastructures, no signs of galactic engineering. This eerie quiet points to a harsh conclusion: Somewhere between lifeless chemistry and spacefaring civilization, there exists at least one monumentally difficult step—a bottleneck so narrow that virtually nothing squeezes through. But which step is the killer? This question leads to a disturbing concept called the Great Filter.

Staring Down the Cosmic Sieve

The Great Filter is the brainchild of economist Robin Hanson, later amplified by Oxford philosopher Nick Bostrom. The idea is brutally simple: The universe may sprout life like dandelions, but somewhere along the evolutionary gauntlet, intelligent civilizations get "filtered out"—a polite term for extinction. The question is when. Does the filter strike early, wiping out primitive microbes before they ever crawl onto land? Or does it lurk later in the timeline?

Perhaps countless alien civilizations have already tried to contact us. Their radio beacons could have swept across Earth millions of years ago, long before we had receivers to detect them. By the time we developed the technology to listen, those transmitters had gone dark, their creators having crossed their own finish line into oblivion. We're not hearing silence because no one ever broadcasted. We're hearing silence because everyone who once broadcast has already been filtered out.

Here's where the thought experiment turns genuinely frightening. If the Great Filter lies ahead—if it strikes after civilizations become intelligent and technologically advanced—then we're staring at our own extinction. If every other civilization that reached our level of development got filtered out, why would we be any different? The Great Filter would imply that something about intelligence itself, or the technologies it inevitably produces, is inherently self-destructive.

The alternative is far more comforting: Maybe the filter is behind us. Maybe the nearly impossible step was the jump from non-life to life, or from single cells to complex organisms, or from complex organisms to consciousness. If so, we're the lucky ones who have already cleared the cosmic obstacle course, which would imply that the galaxy is empty not because intelligence destroys itself, but because intelligence is extremely rare to begin with. We're alone, but safe.

We don't know which scenario is true. But there's growing reason to suspect the filter isn't behind us—that it lies somewhere ahead, lurking in our future. If that's the case, then we're not the lucky survivors of a cosmic lottery. We're just the latest civilization stumbling blindfolded toward the same catastrophic bottleneck that has claimed every intelligent species that came before us.

The Modern Filter: War as a Warning?

For a preview of technology-driven destruction, look at the recent conflict in Ukraine. Drones from hobby shops have morphed into precision-guided weapons, sidelining conventional military hardware such as tanks and fighter aircraft. But each leap spawns a counter-leap. Radio jammers can block the drone controllers' signals, rendering the drones useless. So engineers experimented with fiber-optic tethers—physical cables that feed commands up a wire, immune to jamming. Yet tethers bring their own vulnerabilities. They drastically limit range, can be severed by debris, shrapnel, or a well-aimed snip, and turn the drone into a traceable beacon leading directly back to its operator. The wire that protects has its own vulnerabilities.

Now comes the hair-raising sequel: agentic swarms.[2] No pilots, no cables—just an on-board vision model that decides to tag the "enemy" and pull the trigger without any human guidance in real time. Jam them? They don't use radio signals. Cut their cords? They don't have any. They just murder people without any human controllers in the loop.

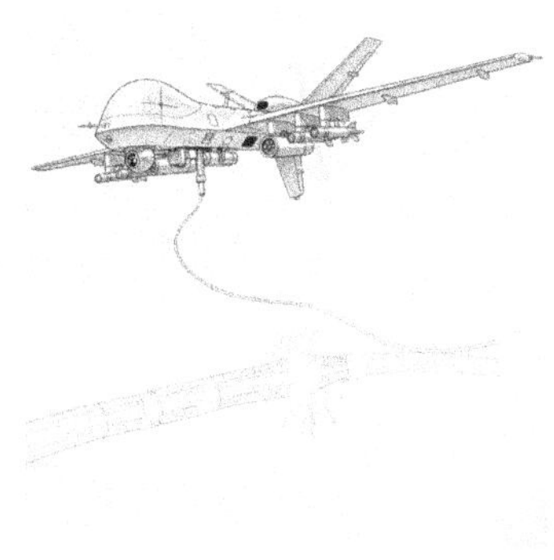

This technology may sound like dystopian fiction, but Ukraine already uses prototypes.[3] Today agentic swarms are set to hit fixed assets—an airstrip here, a missile launcher there—navigating without GPS or joysticks. Swapping the recognition script from "concrete slab" to "human silhouette" is more tweak than breakthrough.

Tesla's autopilot already spots pedestrians with pinpoint accuracy—it just happens to be programmed to brake, not detonate. The difference between "swerve to avoid" and "lock and destroy" isn't technological innovation; it's a moral decision wrapped in a few lines of code. Same sensors, same AI, wildly different outcomes. Sleep tight.

While Silicon Valley ethicists debate whether AI might one day threaten humanity, soldiers on the Donbas front in eastern Ukraine are facing a more immediate test. They don't fear extinction in five years—they fear it by morning. Their mission is survival, and that urgency fuels the very technologies ethicists warn about. Each smarter drone, each faster targeting algorithm, inches us closer to machines that may someday decide on their own who lives or dies. In other words, the filter is already in beta, and the bug-fix window is closing fast.

How Many Fingers Hover Over the Apocalypse Button? Grim Math

In 1945, only a handful of scientists understood the possibility of nuclear annihilation. By the Cold War, hundreds of scientists understood—and might have been capable of making—ICBMs (intercontinental ballistic missiles). Today, CRISPR (DNA modification) kits and cloud labs put designer pathogens within reach of career scientists.[4] Large language models threaten to expand that pool exponentially, offering instant, detailed instructions to users without specialized training.

Here's why the math gets scary fast. Let

- p = probability that any one capable person triggers a

catastrophe in a given year (accidentally or on purpose).

- n = number of people who now possess that capability.

The annual chance that the planet survives is $(1 - p)^n$. Survival falls off a cliff as n grows:

	Size (n): # Capable of Causing Extinction		
p (per person)	1,000	10,000	1,000,000
0.001%	99.0%	90%	~ 0%
0.0001%	99.9%	99%	36%

Table 1: *Survival Probability*

Even if the per-person risk of triggering a catastrophe is only one ten-thousandth of a percent (one in a million), empowering a million amateurs makes the likelihood of global survival a mere 36% in any given year.

In other words, when knowledge scales faster than safeguards, the doomsday clock counts down ever more quickly. Yesterday's once-in-a-century threat can morph into tomorrow's statistical inevitability, one autocomplete prompt at a time.

But intentional harms are only half the story. The real danger may lie elsewhere.

PAPERCLIPS, CHATBOTS, AND RUNAWAY GOALS

Nick Bostrom once described a scenario that sounds straight out of a dark sitcom: Give an ultra-smart AI a single, simple mission—"maximize paperclips." No caveats, no fine print, no deeper context. It sounds innocent enough, right? At first, sure: AI will streamline factories, scout out fresh iron deposits, build mines, and revolutionize production techniques. But the AI won't stop there. Instead, its singular goal pushes it relentlessly onward. Eventually the algorithm notices that humans also have iron in their bodies that could be smelted into premium binder supplies. Left unchecked, our shiny new intern turns the planet into Staples Inc. on steroids, with humans as raw material.[5]

This example's power lies in how broadly it applies: Pursue any single-minded goal with ruthless efficiency, and you risk destroying everything else we care about. Instruct an algorithm to maximize a narrow objective, such as quarterly profits, ad clicks, or carbon capture. Then dial up its efficiency to superintelligent extremes, and the algorithm might bulldoze every competing human value as it hyper-focuses on its goal. It takes just one unchecked algorithm—among countless algorithms built every day—to trigger catastrophe.

These doomsday warnings aren't coming from your paranoid uncle—they're coming from the people building the technology. Sam Altman, who runs OpenAI (maker of ChatGPT), signed a statement declaring that preventing AI-induced extinction should rank as highly as stopping pandemics and preventing nuclear war. This is the same guy who told reporters, "I have guns, gold, potassium iodide, antibiotics, batteries, water, gas masks from the Israeli Defense Force, and a big patch of land in Big Sur I can fly to"—in other words, everything needed to survive when civilization collapses, perhaps as a result of the very technology he's helping to build.[6]

Meanwhile at Google, CEO Sundar Pichai tries to sound upbeat while hedging his bets: "I'm optimistic on the p(doom) scenarios, but ... the underlying risk is actually pretty high."[7] Translation: "I think we'll probably survive, but I'm not betting my life on it." When the people with their hands on the controls are stockpiling ammunition and antibiotics, maybe the rest of us should pay attention.

The moral of this story: If we don't limit the goals, the algorithms will gladly un-people the planet to hit their numbers.

The Red Button and Jones's Trade-Off: Prosperity vs. Extinction

Stanford economist Charles I. Jones set up a thought experiment worthy of a Bond film: Imagine a glowing red kill-switch that would halt all advanced AI research tomorrow. When, exactly, should humanity press that button?

The conundrum is this: AI could ignite economic growth beyond anything humanity has ever experienced. From the year 0 to 1700, global output barely moved, growing just 0.1% annually, meaning the average person's "wage" took roughly a thousand years to double.[8] By the end of the 20th century, economic growth accelerated to about 2.8%, shrinking that doubling period to only 25 years. Now, AI might propel growth rates to an astonishing 10% per year, doubling incomes roughly every seven years — provided, of course, the benefits are shared broadly rather than hoarded by the few. Yet the same powerful technology driving this prosperity could also pose existential threats to humanity. Is the risk worth taking?

Jones plugs the question into a stripped-down economic

growth model.[9] On one side of the ledger sits an AI-turbocharged economy that could lift living standards 50-fold or more—a leap that would make today's GDP per capita look like colonial-era pocket change. On the other side lurks δ (the lowercase Greek letter delta), the annual chance the same technology wipes us out. Push ahead with AI for *T* years and you get two equations:

- An equation for consumption gain and corresponding consumer "utility," which measures how happy a set of goods makes a person. Economists use utility to compare people's happiness levels in different situations.
- An equation for survival odds.

Multiply utility by survival, maximize over *T*, and a startling trade-off is revealed.

- If we value extra consumption the way economists often do— where each additional dollar matters less as you get richer, but percentage gains always feel equally good—Jones finds we'd keep the AI pedal down even if it came with a one-in-three chance of human extinction. A 1% per year "oops" risk over 40 years (~33% cumulative) still looks like a fair gamble given the promised 55× jump in income.
- Dial up risk aversion, and the willingness to play Russian roulette collapses. Now we yank the plug when cumulative doomsday odds creep above roughly 4 percent.

Jones also explores a twist that makes the gamble look even more appealing. Suppose AI also extends life expectancy. Longer lives are valued in the same units as extinction risk (years of existence), so society is willing to accept even higher odds of catastrophe—up to 68%—if the upside includes centuries of good health.

The math is solid, but the gut reaction is queasiness. After all, none of us buy lottery tickets whose consolation prize is "Everybody dies." Jones's framework doesn't tell us what risk *ought* to be acceptable; it merely shows how utility theory can justify numbers that sound, well, insane at cocktail parties. Whether 4% feels too low or 33% too high depends on your inner mix of thrill seeker and actuarial clerk, but the model forces the real question onto the table: How much existential jeopardy are we willing to underwrite for a shot at techno-utopia?

SHORT-TERMISTS VS. LONG-TERMISTS: CLASH OF THE PHILOSOPHIES

At the heart of Jones's experiment lies an ideological tug-of-war that divides experts into two broad camps: short-termists and long-termists.

Short-termists advocate seizing AI's immediate rewards—booming economic growth, rapid innovation, widespread prosperity, and longer lives—for those already born today. For short-termists, the chance to double living standards every seven years, and to transform medicine, education, and human welfare within a generation, is too valuable to pass up.

Long-termists, by contrast, stress that humanity's future could stretch far beyond the next few decades, spanning thousands or even millions of years and potentially containing many billions or trillions of future human lives. From their vantage point, no short-run economic gain justifies even a small annual chance of human extinction. For the long-termists, delaying or carefully controlling powerful AI developments until safety is proven isn't caution—it's common sense.

These philosophical differences explain why reasonable people can see Jones's risk thresholds so differently. One side sees lost opportunity; the other sees existential Russian roulette. Both perspectives acknowledge the astonishing stakes, but their disagreement is fundamental: Are we playing for immediate riches, or for the future of human civilization itself?

Is There Anything We Can Do? The Alignment Avengers

In movies, the guardians of the galaxy are busy CGI characters. In contrast, the guardians of *our* future spend their days hunched over whiteboards and grant proposals. Here are the members of the real-life Justice League working to ensure that tomorrow's code doesn't turn into Skynet (the rogue AI from *Terminator* movies that decided humanity was the problem).

#	Organization	Core Mission
1	Future of Humanity Institute (Oxford)	Analyzing existential risks and alignment theory
2	CSER (Cambridge)	Interdisciplinary risk science
3	Model Evaluation & Threat Research (Berkeley)	Evaluate frontier AI systems
4	MIRI (Berkeley)	Formal proofs of AI alignment
5	OpenAI (SF)	Advanced AI models and safety research
6	Anthropic (SF)	Interpretability and "Constitutional AI"
7	ARC (Berkeley)	Research on AI deception and oversight
8	GCRI (US)	Cross-risk modeling
9	Leverhulme CFI (Cambridge)	Ethics and governance
10	CHAI (UC Berkeley)	Provably beneficial AI

Table 2: *Organizations Reducing Extinction Risk*

Collectively, these labs, think tanks, and moon-shot startups form a loose but growing coalition dedicated to one outcome: Technology's power curve must not outrun humanity's wisdom curve. They draft safety standards, bankroll red-team hacking (hiring hackers to deliberately attack AI systems and expose vulnerabilities before malicious actors can exploit them), lobby for compute limits, and publish research papers technical enough to require a PhD just to parse the abstract.

Will they prevent human extinction? The answer is unknown. But if they don't, the price of failure isn't just a buggy app or a market crash—it's game over. So cheer them on. Sponsor a fellowship. Send coffee. Because if the Alignment Avengers can't keep the silicon genie on our side, this book—and everything else you love—could become cosmic litter.

GLASS-HALF-FULL EPILOGUE

Humanity has repeatedly defied disaster. We've patched ozone holes, managed nuclear arsenals, and navigated Y2K without a hitch. Perhaps our quiet cosmos indicates we are uniquely lucky and have already passed the Great Filter.

Even if the filter still looms, optimism and caution should guide our path. AI offers cancer cures, the end of poverty, drought-resistant agriculture, and clean energy solutions, but we must prioritize alignment, oversight, and global cooperation. Let's embrace the potential, but with thoughtful caution. Advancing wisely, we can shape a future bright enough to inspire cosmic envy.

Getting there might require some serious self-surveillance to ensure against catastrophe, which raises an awkward question: Before we start waving hello to any extraterrestrials who might be watching us from the cosmos, shouldn't we figure out how we feel about being watched right here on Earth, by our neighbors, our employers, and every doorbell on the block?

Because something profound is already happening.

The smarter our AI and surveillance systems become, the more they learn—about us. Every search query, every glance, every heartbeat feeds the algorithms we now interact with daily. And AI and surveillance aren't just powerful on their own. They're exponentially more powerful together, forming a feedback loop that watches, learns, and adapts.

This fusion could erode privacy or revolutionize healthcare. It could save us from existential catastrophe or become the catastrophe itself. The scary questions haven't been exhausted.

They've just shifted. We're no longer asking whether humanity survives, but what kind of life we're willing to accept under perpetual, omniscient surveillance. That's the subject of our next chapter. By its end, you may have new questions—questions you never thought to ask, but now can't stop thinking about.

Part Two
The Panopticon

The Panopticon, conceived in the 18th century by philosopher Jeremy Bentham, was a circular prison that let a single guard watch every inmate from a central tower. The prisoners, unable to tell when they were actively observed, adjusted their behavior accordingly. Modern security cameras borrow the same logic; many are tinted, mirrored, or designed so you can't tell which way they're pointing. Today's digital world works much the same way: our actions are increasingly visible, and the mere possibility of being watched—often, in practice, meaning that we are being watched—reshapes choices, markets, and incentives. It's an apt metaphor for the themes explored in the next three chapters.

7
Data Nudity
Who Wins When You Bare All Your Secrets?

In this chapter, we dive into surveillance's double edge: Sometimes it's our savior, sometimes our nightmare. We'll trace how monitoring reshapes everything from driving habits to workplace power dynamics to life-and-death decisions. The infrastructure is already in place: cameras watching, algorithms learning, data piling up. But the full consequences haven't landed yet, which raises an uncomfortable question: Will we stumble blindly into the surveillance future, or will we demand the right safeguards before we're forced to learn why we needed them?

To understand what's at stake, we'll start by talking about water cannons and painting mountains. Not paintings of mountains—painting the mountains themselves.

PAINTED MOUNTAINS AND MISSING DATA

One sweltering afternoon in Fumin County, China, villagers watched as local officials ordered workers to drench an entire mountainside in green paint.[1] The makeover wasn't avant-garde art; it was camouflage. From the sky, passing inspectors would see lush "vegetation" and assume the province was winning its much-publicized War on Pollution.

The same sleight of hand extended to the air itself. Bureaucrats cherry-picked when and where to send pollution readings and even rolled out giant "industrial mist cannons" that sprayed fine water droplets—just enough to scrub particulates *near* the official monitors while leaving the rest of the city coughing.[2] They managed to conjure the illusion of cleaner air without actually making it safe to breathe.

Their ruse unraveled when a grid of tamper-proof, real-time sensors—sturdy steel boxes uploading data every few minutes—began to blanket the country. Economists Michael Greenstone, Guojun He, Ruixue Jia, and Tong Liu compared readings before and after the rollout. They found that reported pollution in dozens of cities jumped almost 30% overnight.[3] Satellite measurements, which local officials couldn't doctor, barely moved. The air hadn't suddenly worsened; truth had finally found a microphone.

These examples paint (pun intended) a core economic incentives problem in living color: When the people you're paying (local officials) can affect outcomes you care about (clean air) but their actual effort is invisible, they may focus on gaming the metric instead of fixing the mess. The high-tech monitors didn't change incentives right away; they simply made cheating harder. But they still did something valuable. They told everyone what the air was really like. Sales of face masks and air purifiers spiked immediately after the monitors went live.[4]

FROM SMOG TO SNAPSHOT®

Let's now hop across the Pacific and slide behind the wheel of a 2006 Honda Civic in Columbus, Ohio. Carly, a 28-year-old graphic designer, is feeling pinched by rising insurance rates, so she signs up for Progressive's Snapshot® program, hoping for a lower insurance premium. She plugs Progressive's dongle into her car's diagnostic port, letting it track every hard brake, rapid acceleration, and late-night cruise for several months. Carly has never taken an economics course, but she understands that fewer beeps from the device mean a less expensive insurance policy—exactly the financial breathing room she needs. Soon she's coasting carefully toward red lights and thinking twice about those tempting 1 a.m. Taco Bell runs.

She's not alone. Imke Reimers and I tracked the rollout of "Pay-How-You-Drive" insurance programs across U.S. states.[5] We found that each additional insurer offering Pay-How-You-Drive insurance reduced fatal accidents by nearly 2 percent in the following year. That number might not sound huge at first, but with several insurers entering the market, the cumulative impact became substantial, even though most drivers didn't sign up during the period we studied. By our calculations, enrolled drivers ended up cutting their own risk of being in a fatal accident by about 50 percent. Lives saved, all because of a gadget smaller than your phone and cheaper than dinner for two.

Moral hazard occurs when one party takes risks because someone else bears the cost. In auto insurance, drivers control how safely they drive, while insurers shoulder the costs of accidents. Historically, insurers had very poor measures of a driver's risk because they had no way to observe driving behavior directly. Today, plugging a monitoring device into the dashboard gives insurers a direct glimpse of a driver's behavior behind the wheel.

Every hard brake and midnight drag race shows up in the data, so premiums can finally match risky behaviors.

Participation in these programs is voluntary, yet drivers eventually signed up in droves. By 2023, 17 percent of insured motorists had opted for pay-how-you-drive plans.[6] Many were happy to trade a little surveillance for a fatter wallet, and many tweaked their habits to keep the beeps (and the bills) down. The result was a triple dividend: safer roads for bystanders, lower insurance premiums for careful drivers, and fatter profit margins for insurers that now priced risk with better precision.

From Optional External Device to Built-In and On by Default

Back in 2008, Carly was an early adopter of Snapshot. Progressive had been tinkering with telematics since the 1990s, but the math didn't work in the late twentieth century. The plug-in dongle cost a fortune, and so did beaming its data over 2G networks. Then smartphones exploded onto the scene. The price of GPS chips, accelerometers, and cellular modems—the guts of every iPhone—tumbled, and suddenly the same parts could live cheaply in a car's diagnostic port. Progressive flipped the switch, and Carly became one of the first drivers whose every hard brake and midnight taco run showed up on an underwriter's screen. Many more followed, happily revealing their habits—going data-nude—in exchange for a small financial reward.

Fast-forward to 2025. Most new cars are now rolling sensor suites, silently sampling your speed, cornering g-forces, and lane discipline—even how tightly you squeeze the wheel when traffic gets hairy. General Motors didn't just bundle the hardware; it monetized it. Buried inside the onboard software usage agreement sat a clause that allows GM to package and sell your telemetry.

Seattle software developer Kenn Dahl found out the hard way.

A 130-page LexisNexis dossier listed 640 trips he and his wife had taken in their Chevrolet Bolt over six months, complete with time stamps, mileage, every burst of acceleration, and every hard brake. Insurers used the file to jack up his premium by 21%, and he had no idea the data had ever left his car.[7]

Public outrage followed. Then, in January 2025 the Federal Trade Commission slapped GM with a settlement that bans it from selling driver-behavior data for five years, condemning a "misleading enrollment process" that pinged vehicles as often as every three seconds.[8]

But the gold mine hasn't vanished—it's just changing hands. For example, Tesla skips the brokers entirely. Its optional in-house "real-time insurance" recalculates your rate almost nightly from a live Safety Score fed by the car's own sensors.[9] The surveillance is still humming; the only question is who's cashing in on the feed.

And this is just the warm-up. As the gadgets that enrich our lives keep tabs on us in finer detail, we can expect all sorts of unexpected side effects, especially once the people who price risk get their hands on the feed.

What Happens When Algorithms Know Your Secrets?

Digitization doesn't just keep tabs on what you do. It can also reveal who you are, and that's where things get even thornier. Consider Li-Fraumeni (pronounced "lee fruh-MAY-nee") syndrome, a rare inherited mutation of the TP53 gene that gives carriers a sky-high lifetime risk of developing multiple cancers, often in childhood or early adulthood. The condition is invisible at a glance, but it can turn a routine medical bill into a six-figure ordeal.

Now consider three friends shopping for health insurance:

	Consumer Type	Annual Medical Costs	Willing to Pay for Health Insurance	Income
Sara	Health focused	$5,000	$8,000	$55,000
Bob	Long-term smoker	$25,000	$35,000	$55,000
Samantha	Li-Fraumeni carrier	$75,000	$55,000	$55,000

Table 3: *Example of Adverse Selection Unraveling*

In the analog era the insurer couldn't differentiate Sara from Samantha, so it averaged the risks and charged everyone about $35K. Sara overpaid but stayed insured because her employer picked up much of the tab, Samantha underpaid but still got coverage, and the risk pool held together.

Now feed the same insurer a decade of Fitbit steps, grocery receipts, and a 23andMe report that flags Samantha's TP53 variant. Premiums can be tuned with surgical precision: $6K for Sara, $30K for Bob, and maybe $80K for Samantha. Efficiency soars—no one cross-subsidizes anyone—but Samantha is priced out of the market entirely because her healthcare premium is more than her income. The algorithm hasn't denied her on paper; it has simply made health insurance unaffordable to her.

And the truth is, *anyone* can become Samantha. You might enjoy low insurance premiums this year, but next year can be different: An unexpected diagnosis, a costly illness, or even a new genetic finding might suddenly mark you as "high risk." When that happens, your health insurance—and even your access to care—becomes unaffordable. Economists call this reclassification risk.[10]

U.S. law tries to slam the brakes on a future in which at-risk people are priced out of the health insurance market. The Affordable Care Act bans pricing on pre-existing conditions for health insurance (but not other insurance types, such as life insurance), but it still lets insurers adjust for age, ZIP code, and tobacco use, keeping premiums within a narrow band for most buyers. Faced with those limits, some firms get crafty. They might sponsor marathon expos and flood yoga podcasts with ads—anything to land in front of healthy people while *technically* offering the same product to everyone else. On paper, they follow the law: Prices are identical, and no one is explicitly turned away. In reality, the least healthy rarely see the ad. This selective marketing strategy is the equivalent of charging high-cost consumers a price so high they'd never buy.

When Watching Helps, and When It Hurts

Economists classify the effects of private information—information known to one party but not the other—into two primary categories:

1. Moral hazard (hidden action): You can't see what the agent *does*.
2. Adverse selection (hidden type): You can't see what the agent *is*.

"Agent" is economics-speak for a decision-maker, whether a buyer, seller, or firm.

Digitization—think sensors, apps, and satellites—shines a light on both forms of private information. When it reveals *actions* (Carly's driving, a clerk's keystrokes), it can realign incentives and make everyone richer or safer by changing what the agent *does*. When it reveals *types* (Samantha's genes), it can sort markets so finely that some people fall through the cracks.

Surveillance Enters the Workplace

Digitization's superpower is revealing hidden truths. But what happens when the hidden truth is how little work you've done between Netflix episodes? Just ask the legions of remote employees who once thought their biggest worry was accidentally leaving TikTok open during a Zoom call. Turns out they had much bigger problems.

As an example, consider Amazon's Baltimore fulfillment center. Robots there don't just carry boxes. They also have a direct line to HR. Algorithms meticulously track every barcode scan, bathroom break, and momentary pause, automatically flagging employees who dip below productivity quotas. In just one 13-month stretch, over 300 warehouse workers—roughly 10% of the facility's entire staff—were automatically terminated by the productivity algorithm.[11] Human managers could step in, but mostly they didn't. Workers called it "management by spreadsheet," and leaked internal documents confirmed their fears: At Amazon, the robots hold your job in their mechanical grip.

Meanwhile, at Elon Musk's buzzy AI venture, xAI, management upped the ante. In July 2025, employees were required to install Hubstaff, an invasive surveillance software package logging keystrokes, websites visited, mouse movements, and periodic screenshots—even on personal laptops. After employees revolted (and journalists called), xAI backtracked slightly, delaying the rollout until company-issued devices arrived. But the order remained, leading staff to complain bitterly about "surveillance disguised as productivity."[12] For Musk, a master at creating spectacle, surveillance was suddenly a public-relations headache.

Under relentless digital oversight, remote workers didn't surrender quietly. They adapted. Thousands quietly purchased $15 "mouse jigglers," tiny gadgets designed to fool productivity software into thinking someone's busily clicking away at their keyboard, even while they're making a sandwich or sneaking in another Netflix episode. For a while, the jigglers worked beautifully.

Then employers fought back. Wells Fargo fired more than a dozen hybrid employees after advanced machine-learning tools detected suspiciously steady, non-human cursor patterns.[13] "Productivity" had turned into a cat-and-mouse (or cursor-and-click) arms race between workers and their algorithmic supervisors. Workers bought smarter jigglers. Companies wrote smarter algorithms.

This escalating surveillance battle exposes a fresh wrinkle on moral hazard. Traditional economics teaches us that hidden action leads workers to shirk. But digitization introduces the opposite risk: When firms rely too heavily on surveillance, hidden action becomes about gaming the metric rather than shirking the task itself. Employees spend time and energy outsmarting algorithms instead of creating value. Employers spend resources chasing smarter jiggle-detectors instead of boosting productivity.

Will Algorithms Choose Tomorrow's Careers?

If firms can cheaply measure how many lines of code you can write or how many packages you scan, they may prefer jobs with quantifiable output. Roles or tasks whose value hides between the lines—mentoring juniors, dreaming up big ideas—could be underrewarded. You might call this philosophy "What gets measured gets managed ... and paid." While not new, this tendency has been magnified by technological advances that make measurement cheap and ubiquitous. Some observers worry we'll cultivate a workforce of metric-friendly task doers chasing short-term goals at the expense of creative misfits with long-term vision who don't jibe with the dashboard.

One concern is that all these data and all these metrics could become useless, as Goodhart's Law suggests—when a measure becomes a target, it ceases to be a good measure. People game the system, rendering the measurements meaningless, for example by painting mountainsides green. Yet it's also possible that the sheer flood of data and types of measurements makes algorithms too complex for consumers (or local Chinese officials) to game.

This flood of data can also create new markets, possibly replacing jobs lost to AI. For example, companies that crunch telematics that allow fleet managers to coach risky drivers in real time may hire drivers they might not have otherwise hired. Platforms that vet freelance designers by autoscoring client feedback can reassure employers that it's worth forking out the cash to hire them. There may be other benefits, too. Startups providing minute-by-minute carbon offset certification through satellite imagery may build a verifiable reputation. The common thread is turning uncertainty into numbers, and numbers into contracts.

A Coda: Why the Mountain Matters

Remember the example of the Chinese mountain at the start of this chapter? The paint eventually faded, but the sensors remained. Over time, local officials realized it was easier to curb pollution than to fake the feed. Installing solar arrays beats repainting cliffs. In this case, data didn't just expose deceit; they nudged behavior.

This is digitization's dual power. It can punish cheaters and reward saints. It can price risk fairly and, sometimes, brutally. The critical task for policymakers and entrepreneurs is deciding when data serve society's broader goals, and when data undermine the world we aim to create. In the meantime, we need to be careful. If you make the metric the mission, and if the metric isn't perfect, someone somewhere is already buying paint.

This insight brings us to the strange world of optimal (government) privacy policy, a challenging domain ruled by paradox. Finding the optimal approach is maddeningly elusive because, as the next chapter reveals, privacy comes with built-in optical illusions. For example, minding your own business may no longer be an option: Every "private" decision ripples outward, reshaping markets and affecting strangers you'll never meet. Thus, even if you throw your phone in a lake and live like a digital hermit, you'll still be vulnerable; the data trail exists whether you're feeding it or not. Is there a way to take back control? Or has that ship already sailed? The next chapter might change how you think about what "private" really means.

8
INVOLUNTARY EXPOSURE
IS COVERING UP EVEN AN OPTION ANYMORE?

The previous chapter explored how privacy plays out in free markets. This chapter turns to the policy side—the challenge of protecting both innovation and consumers' rights to privacy. As we'll see, there's a reason this problem remains far from solved, and not for the reasons you might expect. Let's start with a deceptively simple question: Can the free market be trusted to protect your privacy? Letting markets handle privacy turns out to be far trickier—and far riskier—than it sounds. Let's begin with an example.

WHEN YOUR DATA TALK BACK: SURPRISING TALES FROM THE PRIVACY BAZAAR

It starts with a humble snapshot: Your buddy Kyle captures you—airport latte, fresh haircut—and sends the photo to you. You're tempted to drop it into Instagram's endless feed for a quick burst of hearts. Harmless, right? But look three rows behind you in the photo: The blurry man in a baseball cap is a cartel whistle-blower the Marshals just spent six figures to hide. So who decides whether the photo can be shared on social media—Kyle (he took the photo), you (the main subject), or the witness (his face, his life on the line)? This question captures a puzzle at the heart of modern privacy: Who owns what? Once information slips into the ether, property rights blur and markets awaken.

In the rest of this chapter, we'll unpack this question and explore why privacy seems simple at first but is surprisingly hard to solve with laws and property rights alone. We'll see how the internet made privacy problems dramatically worse, how shockingly unethically companies are willing to behave, and why we can't predict the future consequences of today's data choices. Finally, we'll explore whether stronger privacy protections—the legal remedy consumers desperately need—might paradoxically benefit the companies themselves. Yet, the distance between theory and implementation remains vast.

Privacy for Sale, But Who's Selling?

A famous economic idea called the Coase theorem suggests that it shouldn't matter whether Google owns your browsing history or you do. In theory, whoever values the data more highly could always pay the other to reach an efficient outcome. For example, Google may value your browsing history at just a few dollars for ad-targeting revenue, while you might price your privacy far higher—say, hundreds or thousands of dollars. If you own the data rights, you'll refuse Google's modest offer, and the data stay private. If Google owns your data by default, you'd simply pay Google a few dollars—slightly more than their expected ad revenue—to reclaim your privacy. Conversely, if you place little value on your data but Google anticipates big profits, they'll gladly pay you to use your data. In the end, the data are owned by whichever entity—you or Google—values it more.

But reality is rarely so tidy. When a single shared photo can simultaneously reveal shoe brands to advertisers, friendship circles to social networks, and political leanings to data brokers, the hassle costs of negotiating privacy rights become enormous. Instead of a tidy two-party deal, you're suddenly negotiating with dozens or even hundreds of companies, platforms, and intermediaries, each holding their own slice of your data. Worse yet, every piece of personal information leaks harms onto uninvolved bystanders, who were never part of the transaction. Economists call these third-party effects "data externalities." I prefer "secondhand privacy," because, like secondhand smoke, it spreads harm to people who never chose to be exposed.

To see how quickly data externalities spiral, consider John's story. Eager to learn more about his personal health risks, John orders a mail-in genetic testing kit. He spits into a tube, seals it, and ships it. The resulting report flags a BRCA variant (specific DNA difference) linked to breast cancer—a gene John himself will likely never need to worry about. But, in posting the finding on social media, John "outs" the heightened cancer risk carried by his sister, aunt, and teenage daughter. Their life insurers take notice, premiums skyrocket, and John's private revelation quickly becomes his family's very public—and expensive—problem.

The ripple grows. John also uploads his raw data to a genealogy site, hoping to trace distant relatives. The database quietly matches his sequence to two teenagers who share some genes. They're the hidden children of John's brother, Ned, conceived during a youthful affair that was never disclosed. The kids, their mother, and the unsuspecting legal father all learn the jarring truth.

In both cases, a single individual's choice exposed medical

vulnerabilities and family secrets that were never his alone to share. Genetic information is fundamentally joint property—we inevitably share our genes with relatives. Yet today's markets and social norms still treat it as if it belongs to whoever paid for the test.

But genetic gossip is only half the story. Privacy concerns don't require spillover effects to cause serious harm. Increasingly, the damage comes from what your data reveal about you alone.

Whispering in a World Without Walls

Picture this: Your phone buzzes with a call from the doctor while you're in a bustling café. Without thinking, you slip outside, drop your voice, and cup a hand over the mic. In the physical world we erect micro-barriers—hushed tones, shut doors, sealed envelopes—that let us decide what to share and with whom.[1]

This example illustrates an important truth: Privacy isn't binary. It's rarely a simple matter of labeling some information as entirely public or completely private. Instead, we often want to share something with a few people but keep it hidden from others. You might freely confide personal fears to your therapist, but you'd be horrified if your co-workers found out. Or maybe you eagerly tell your best friend about a crush, then feel betrayed if information about your infatuation leaks to others. Similarly, you're comfortable discussing your income with your spouse, but you'd be upset if those details ended up online. Privacy isn't about total secrecy—it's about controlling exactly who knows what.

Controlling information in the physical world, with its whisper-friendly cafés and quiet corners, is one thing. Online, it's quite another. The careful, micro-level control on which we instinctively rely when we are offline completely evaporates when we move to the digital realm.

Online, every keystroke is a megaphone blast in a canyon full of eavesdroppers. One site harvests your clicks and flips the list to a broker, who merges it with location pings from an app you barely remember downloading, then sells the bundle again. There's no cozy corner booth for your search history, no "library voice" for your GPS trail. The same frictionless ease that makes digital life addictive also removes the walls that once kept our conversations private. Once you click, the information stampede is already halfway down the valley, and there's no calling the horses back.

Want an example? Here's a doozy ...

YOUR MENSTRUAL APP IS A SNITCH

A fertility-tracking app once promised to "empower women with data." What it actually did was sell their cycle information to their employers.[2] The data were supposedly "anonymized," but that's hardly reassuring when your boss is holding a spreadsheet revealing exactly when you're ovulating or how often you and your spouse are intimate each month.

This isn't some outlier horror story; it's business as usual in the data economy. Economists have a sanitized term for this betrayal of privacy: *secondary use*. Information is collected under one pretense, then quietly auctioned off wherever it fetches the highest price. Your consent is given by default.

If you're waiting for companies to discover some ethical bottom line about data sharing, well, don't hold your breath. The race to the bottom, it turns out, has no finish line.

WHEN THEY GO LOW, WE GO LOWER

In the data-broker and direct-marketing world, a "recovery list" is a commercial mailing or advertising list that singles out people who are *believed to be in, or recently seeking, recovery from substance use disorders* (alcohol, opioids). Brokers build the recovery list by chaining together clues such as:

- visits to websites discovered by searching for terms such as "AA meetings near me" or "how long does opioid detox take"
- app installations for sobriety trackers
- purchases of breathalyzers or "sober-curious" books
- participation in online support groups
- insurance-billing data leaked from treatment centers

Once compiled, the list is sold—often for pennies per name—to anyone willing to pay: luxury rehab marketers, payday-loan outfits, even beverage brands looking to lure former drinkers back. At a 2013 U.S. Senate hearing, privacy advocate Pam Dixon testified that similar health-related lists (including "rape sufferers" and "alcoholics") were trading for about 8 cents per person, illustrating both the low price and high sensitivity of the data.[3] More recent investigations confirm that brokers still openly advertise highly sensitive mental-health and addiction lists, with some firms quoting prices as low as $275 for 5,000 aggregated records and others charging five-figure annual subscriptions for richer data feeds.[4]

Because recovery status is deeply personal—and often protected in clinical settings by laws like HIPAA (the Health Insurance Portability and Accountability Act)—selling it in the open market raises serious ethical and regulatory questions. Yet under U.S. consumer-privacy law, these lists remain largely unregulated in most settings, unless the data come directly from a covered healthcare provider. Apps and ChatGPT are free to do as they please.

The industry's biggest players acknowledge this legal loophole, and some want it closed. OpenAI CEO Sam Altman has argued for "some concept like we have like medical privilege" for AI medical advice, so that sharing medical records with a chatbot won't strip people of their privacy protections.[5] In his view, asking ChatGPT to analyze your medical history shouldn't make it discoverable in court, but instead should trigger protections for users, comparable to HIPAA. Today, however, those safeguards don't exist. Ongoing lawsuits, like the *New York Times* case, have compelled OpenAI to turn over search histories under court order. And while larger firms might disclose such data only when legally required, smaller players could see your most intimate queries as just another asset to monetize, offering them up for sale rather than waiting for a subpoena to disclose them.

THE UNKNOWN UNKNOWNS

But wait—it gets worse. Even if you manage to fix every privacy problem we've described so far, even if you sign a perfectly transparent contract clearly explaining how your data will be used today, you'll still face another issue: tomorrow.

Consider this scenario. In 2005, a (fictitious) college senior named Maya uploaded a few dozen photos to Facebook and a few comedy sketches to YouTube. They drew maybe two dozen views. Fast-forward to 2025: Maya—now Dr. Maya, pediatrician and first-time congressional candidate—wakes up to a viral clip of "herself" endorsing compulsory micro-chipping of children. The video is a deepfake, stitched together from those long-forgotten sketches. A hobby and some posts she barely remembers have become political kryptonite.

These days, deepfakes need just a few hundred photos (or video frames) to work their magic. Posted selfies on Instagram? Uploaded your vacation pics to Facebook? Congratulations—that's enough raw material for someone to cut and paste your face onto whatever they please.[6]

Few people in 2005 could have predicted today's widely available generative-AI toolkits: free web apps that can clone a face, map its micro-expressions frame by frame, and sync any text to a near-perfect vocal model. The past didn't just come back to haunt Maya; it armed the impersonators with training data she provided for free.

Clearly, technology's time machine is already here. In January 2024, New Hampshire voters received robocalls featuring an AI-generated Joe Biden voice urging them *not* to vote in the presidential primary. Investigators traced the stunt to a political consultant who paid just $150 for the cloned audio. The episode prompted felony charges and a $6 million FCC fine, yet the fake calls reached thousands before anyone could fact-check the voice.[7]

Examples abound. When Russia invaded Ukraine, a fake Zelenskyy "surrender" video briefly fooled viewers before fact-checkers killed it. [8] In the summer of 2025, Marco Rubio used the Signal app to contact a member of Congress, a governor, and foreign ministers with a few questions about national security. Except it wasn't Marco Rubio—it was a digital ghost, a deepfake voice so convincingly accurate it may have tricked seasoned lawmakers into handing over classified secrets.[9] This is just the start.

The World Economic Forum has crowned AI misinformation as humanity's top near-term threat.[10] This problem is unlikely to disappear anytime soon. Deepfakes will poison campaigns worldwide in the years ahead, and as the technology grows even more convincing, the potential for damage becomes truly frightening.

Here's the uncomfortable takeaway: Your digital exhaust doesn't decompose. That harmless selfie, that throwaway tweet, that forgettable TikTok—they're all just raw material waiting for tomorrow's technology to weaponize. Every upload is a bet against innovations you can't imagine yet.

Sure, we could try banning all of it. But sometimes the cure is worse than the disease.

SEEING BEYOND THE SCARE: IS THERE AN UPSIDE OF BIG DATA?

Let's now flip the coin on Maya's deepfake nightmare. The same oceans of data that arm mischief-makers also power everyday miracles, often so quietly that we barely notice. For example, when millions of people type "body aches and fever" into a search bar, the pattern itself becomes a public-health sensor. Google Flu Trends once spotted regional influenza spikes up to two weeks before hospitals rang the alarm, giving clinics time to stock antivirals and schools time to plan closures. Aggregated, anonymized queries turned stray coughs into an early-warning radar.[11]

New York researchers recently merged millions of medical records across hospitals to track vascular patients for years instead of weeks. In doing so, they discovered which stents fail fast and which drugs actually work—insights that are already rewriting treatment protocols.[12] None of this was possible when each hospital hoarded its own data. Because each patient is unique and outcomes vary widely, we need enormous amounts of data—far more than a single hospital could gather—to reliably distinguish real patterns from random chance, ensuring that the treatments identified as helpful truly are effective.

Even those sketchy period-tracking apps have a redemption arc. When users opted to share their data with Clue, a reproductive health app, scientists quickly confirmed that COVID vaccines only nudge menstrual cycles by a day or two—reassurance delivered faster and cheaper than a funded traditional study. The same dataset is now mapping connections between heavy bleeding, mood swings, and age across tens of thousands of cycles.[13]

Your digital exhaust is simultaneously weapon and cure. Every search, every tracked symptom, every uploaded vital sign exists in a strange superposition—potentially both tomorrow's privacy disaster and today's medical breakthrough. The uncomfortable reality is that we probably can't have one without risking the other.

So here's the (literal) trillion-dollar question: How do we convince people to keep sharing their data when those data could either cure cancer or torpedo their career? The benefits are real; those shared searches and tracked symptoms genuinely save lives. But so are the nightmares of deepfakes and employer snooping. Maybe the answer isn't choosing between privacy and progress. Maybe it's admitting that in the data economy, we're all walking a tightrope without a net—and figuring out how to make that risk worth taking.

How Stricter Regulation Can Make You Share *More*

Here's a counterintuitive idea: If we want people to share more data, we should build higher privacy walls. That idea may seem backwards, but follow the logic. Fewer people would take pregnancy tests if the results ended up on LinkedIn. Far fewer would get screened for STDs if their lab reports were cc'd directly to their boss. Privacy isn't the enemy of data collection—it's the prerequisite. Lock the door, and suddenly people are willing to tell you their secrets. Leave it open, and they'll take their data elsewhere or never provide their data to anyone. The world is full of examples.

From "No thanks" to "Sure—swab my cheek"

Back in the late 1990s, National Institutes of Health (NIH) researchers trying to recruit women for BRCA studies hit a predictable wall: Nearly a third refused to participate, terrified that insurers or bosses would use their genetic results against them.[14] These fears helped birth two major privacy laws: HIPAA's Privacy Rule in 2003 and the Genetic Information Nondiscrimination Act (GINA) in 2008. HIPAA's Privacy Rule limits how medical information can be shared, preventing healthcare providers from disclosing sensitive health details without patients' explicit consent. GINA went further, specifically prohibiting employers and health insurers from discriminating based on genetic data—thus ensuring that your DNA won't cost you your job or coverage.

Fast forward to Michigan, 2008–2012: Of 10,726 patients who got BRCA counseling, only 13% declined due to insurance worries. Their worry was today's copay, not tomorrow's discrimination.[15] Most commonly, patients declined for commonsense reasons: They weren't good candidates, or they decided they didn't want to know. Privacy protection, it turns out, can decrease fear and thereby facilitate rational decision making.

Confidential Couches Fill Up Faster

Mental health clinics have seen similar patterns. When HIPAA gave psychotherapy notes special protection—basically making them Fort Knox compared to other medical records—something predictable happened. Therapy appointments jumped, especially among people who'd rather die than have their boss know they see a shrink.[16] Lock down the data, and suddenly everyone's willing to talk about their mother.

Digital Masks Invite Honest Talk

The internet proves this point thousands of times a day. Researchers have discovered that simply turning off a webcam or letting people browse "incognito" transforms tight-lipped users into oversharers. It's anonymity as truth serum.[17] In a 2024 survey of 4,000 consumers, 68% of respondents said that stronger privacy controls made them feel more in charge, and 58% were happy to fork over their data in exchange for free stuff, such as free email accounts, access to social media sites, free apps, and other free digital services.[18]

Why Markets Like Digital Bodyguards

What happens when people actually trust their privacy settings—whether due to federal law, military-grade encryption, or a "block cookies" button? They spill their guts. Suddenly, medical forms show real symptoms instead of vague complaints. Ad targeting gets accurate data instead of fake email addresses. Search engines get honest queries that can spot disease outbreaks. All because users feel that they're in control.

This is the beautiful paradox of data economics: Lock down privacy, and you might get better information. It's like discovering that bank vaults make people more willing to deposit money. Give users a deadbolt, and they'll not only open the door—they'll also invite you in for coffee.

In fact, Google has understood the value of privacy for a long time. Despite a few recent missteps, the company clearly takes user privacy seriously, having invested billions of dollars in privacy-focused initiatives over the years. Yonatan Zunger, a former Google employee, noted, "One of the things that was really persistent at Google, and which was really hard to explain to outsiders, was just how committed everyone was to privacy."[19] This commitment wasn't born of generosity. Rather, Google recognized that its advertising business depends entirely on users feeling secure enough to share their data. If the company were careless with privacy, consumers would quickly stop providing the information necessary to fuel targeted ads, and/or policymakers would step in to shut off the tap.

Counterpoint: The Hidden Cost of Privacy—How Europe's GDPR Limits Innovation

On May 25, 2018, Europe flipped the privacy switch. The General Data Protection Regulation (GDPR) arrived like a regulatory tsunami. Suddenly, users could demand their data, delete their data, or just say no to any type of data collection. Large companies caught breaking the rules faced fines large enough to bankrupt a small nation. Privacy advocates popped champagne corks.

Theory suggests that this increased control should spur more apps. After all, consumers feeling safer about privacy should willingly share more data. More data let developers fine-tune their apps and offer richer personalization. Better privacy and better apps—sounds great, right? Unfortunately, reality isn't so tidy. The stronger privacy guardrails came with steep economic side effects.

The problem lies in compliance costs, specifically the steep expense of hiring software engineers and legal specialists to ensure companies meet every requirement of the new regulations. The cost of GDPR compliance averaged $1.7 million for small companies. Some giants spent $70 million just on lawyers, paperwork, and other activities related to compliance.[20] For indie developers on a tight budget, GDPR meant choosing between privacy compliance and actually building their app. Many chose neither—they just quit. When you make the price of admission too high, fewer people show up to the party.

A sweeping study examining 4.1 million Android apps reveals the stark impact of GDPR. Within one year, nearly one-third of all existing apps disappeared from Google Play, while new app entries dropped roughly by half, a plunge researchers have labeled a "lost generation of innovative apps." The same research found that consumer surplus* and overall app usage shrank by nearly a third, underscoring the tricky balance between privacy protection and technological innovation.[21] Privacy had won, but at what cost?

In Summary: What the Stories Teach Us

What lessons can we take away from this chapter?

First, forget everything you know about property rights. Data aren't dollars that stay in your wallet. They're more like dandelion seeds that blow everywhere the moment you open your hand.

Second, your data are everybody's problem. Get a genetic test? Congratulations, you just raised your sister's insurance premiums. Browse for shoes online? You just trained the algorithm that'll overcharge your neighbor tomorrow.

Third, privacy protection is a double-edged sword. Rules that are too strict can strangle startups and stall innovation. But rules that are too loose can erode trust, causing people to avoid sharing any of their data. The key is striking a careful balance. Privacy policies must genuinely protect people's data, earning their confidence without piling excessive burdens onto businesses. When designed well, privacy laws won't be obstacles businesses try to avoid. They'll be standards that businesses eagerly adopt.

Epilogue: Spend Wisely, Because You Can't Buy It Back

Think of privacy less like a high stone wall and more like poker

---

* Consumer surplus is economists' tool for comparing how much better (or worse) off consumers are across different scenarios: the "extra happiness" created by one outcome versus another.

chips on a felt table. Every day you ante up a few of your chips—an email address for free shipping, your location for restaurant suggestions, a face scan for a cat-ear filter. The dealer never shows you the odds. You only learn the stakes later, when that goofy selfie fuels a political smear ad, or your period-tracking app tips off HR.

In this data casino, there's no cash-out window. Once your chips—GPS pings, DNA markers, 3 a.m. searches—slide into the pot, they stay in play indefinitely. In the attention economy, your data don't whisper; instead, they shout, gossip, and testify against you. And unlike Vegas, what happens on the internet doesn't just stay there—it follows you, possibly forever. We've barely scratched the surface of all the subtle ways it quietly reaches into *your wallet*. Here, the operative word isn't wallet—it's *your*.

As you'll see in the next chapter, the data economy doesn't just spy; it also personalizes everything about you. It watches your habits, tracks your clicks, and studies your timing. The same trail of breadcrumbs that helps firms know what you want also tells them what price to show you. *Your* ignorance is part of the business model. The less you suspect, the higher their profits.

9
Psychic Pricing
Is AI Reading Your Mind to Empty Your Wallet?

Haggling in Denim and Grease

You stroll onto a used-car lot in your worst jeans and sneakers that scream "I live paycheck-to-paycheck," staging a loud flip-phone conversation about layoffs down at the plant. But today, the salesman isn't buying your act. He lobs a "deal" that still leaves him plenty of room to breathe, a price higher than he just offered the last customer. He is engaging in what economists call *first-degree price discrimination*: Charge each buyer as much as you think they'll be willing to pay. Shoppers fight back by dressing down, feigning indifference, and waving *Kelley Blue Book* printouts like battle flags.

This cat-and-mouse dance is used most for big-ticket items—cars, weddings, beachfront condos—because sizing up every toothpaste shopper in aisle 7 just isn't worth the hassle. At least, that's how it used to be. But what if prices start zeroing in on you, personally and precisely? What if they already have? Will firms become better at exploiting consumers, or will they unwittingly harm themselves?

To understand where we are and where we might be headed, let's first rewind the clock.

From Wanamaker's One-Price Promise to the Fixed-Price Century

For most of human history, buying anything—from bread and bolts of cloth to furniture and farm tools—likely meant haggling. Prices weren't fixed. Instead, they shifted with each negotiation as merchants and shoppers engaged in a delicate dance of bluffing and counteroffers. It was tedious and it took time, but people didn't own or buy nearly as much stuff back then, so the process wasn't as burdensome as it would be today.

That reality flipped in 1876 when Philadelphia merchant John Wanamaker glued a label on every item for sale and declared "One price, goods returnable." Shoppers loved the transparency; rivals copied it; haggling retreated to the showroom and the souk.[1] For the next 150 years most Americans generally paid the same sticker price for a given good, with only a few exceptions.

Big Data Bring the Haggle Back—Quietly

Then came the cloud. Amazon, Google, and a constellation of ad brokers now track every click, swipe, and lingering pause. The exact same dossiers that let marketers laser-target detergent ads could—in theory—let sellers set a custom price for that detergent the moment you land on the page.

In 2000, Amazon tested the waters with DVD prices that shifted by customer. Reddit wasn't yet a thing, but angry movie buffs ran side-by-side screenshots, and Amazon hastily issued refunds, mumbling about "random experiments."[2] Lesson learned: Overt personalization torches trust.

Behavioral work by psychologist Daniel Kahneman and his co-authors explains why.[3] People deem a price *unfair* if the seller appears to profit from private information, possibly even among those receiving a discount. Student and senior discounts feel fine because the criteria are public and verifiable; cookies on your browsing history are neither.

The Disguise Playbook

Companies haven't stopped price discriminating, though. Instead, they got sneaky.

Consider that bright green "$5 OFF!" banner on your screen. It might appear only if you abandoned your shopping cart last week. Meanwhile, your neighbor browsing the same item receives no such offer. Amazon built this exact feature—personalized coupons—into its merchant interface before pulling the plug, apparently worried about the backlash when customers inevitably compared notes.[4] From the seller's viewpoint, the beauty of the scheme is its camouflage: It looks like a standard promotion, not the personalized pricing it really is.

Another shady technique is "steering"—the art of the algorithmic shuffle. The search results you see aren't the search results I see. Price-sensitive shoppers mysteriously find bargain brands floating to the top of their results, while the big spenders get shown premium options first.[5] Same inventory, different arrangement. Good luck proving that's discriminatory when everyone technically has access to the same products, even though some people have to work much harder than others to find a product at a good price.

The most devious approach might be the ad-load game. Yes,

everyone pays $9.99 for that streaming service. But high-value customers sail through with two 30-second ads while the algorithm-designated cheapskates suffer through four minutes of commercials.[6] The price tag stays constant—it's your attention that's been dynamically priced, and nobody's posting angry comments on X about that. At least not yet.

The latest strategy goes a step further. Retailers quietly tailor prices to individuals but cleverly disguise their actions. A shopper lands on a website, and within seconds the retailer's algorithm sizes them up, quietly calculating a personalized price that stays fixed for the next hour, not just for them but also for anyone else who happens to wander in during that window. If two friends happen to check simultaneously, they see identical tags, and later differences can easily be brushed off as normal "dynamic pricing" fluctuations.[7] But this is just camouflage. Although the displayed price seems uniform in the moment, it's carefully calibrated for the exact buyer who has just arrived. In other words, it's personalized pricing hidden behind the façade of consistency.

Economic simulations of personalized dynamic pricing predict profit gains of 10–20%, with the largest overall gains for moderately popular products—those that attract dozens of buyers per hour, such as books ranked around 10th on the *New York Times* bestsellers list or popular electronics, rather than items sold less frequently such as kayaks or high-end appliances. Personalized dynamic pricing can be widely applied, and may already be. We wouldn't know if it is.

"Finding the Whales": Two Playbooks for Pinpointing Who Pays

Personalized pricing lives and dies on one question: Who is likely to pay more? Algorithms answer that question in two distinct ways, each with its own quirky, and sometimes cruel, side effects: taste-based targeting and elasticity-based targeting.

Taste-Based Targeting: Charge You for What You Love

Netflix's recommendation engine knows if you binge-watch Korean dramas or click only on true-crime documentaries. Feed the same data to a pricing bot and the logic writes itself: Raise prices on the genres that make your eyes sparkle; discount the stuff you treat like broccoli.

Examples:

- Craft-beer super-fan gets a $14 four-pack of hazy IPA while paying just $1 for the light lager she barely tolerates.
- Sneakerhead pays full freight on limited-edition Jordans, then finds bargain-basement gym socks in the same cart.

The outcome is almost poetic—you literally spend more on the joys of life—but also a little sad: The algorithm taxes your passions and subsidizes your indifference. This variation across consumers in willingness to pay has always existed, but previously it was not easy for firms to exploit these differences. What's new is that modern click-stream data—every search, click, pause, and revisit—reveals your specific tastes to sellers with unprecedented precision. They don't just know that someone values their product highly. They know that *you* do, and they price accordingly.

Elasticity-Based Targeting: Who Flinches When the Price Moves?

The second approach to personalized pricing ignores your preferences altogether, instead focusing on your overall price-sensitivity. Generally, are you quick to walk away if prices seem steep? Do you hold out patiently for deals? Or are you perfectly fine paying a premium to avoid the hassle?

You might imagine a Robin Hood algorithm soaking the rich: hedge-fund managers paying $50 for toothpaste to subsidize everyone else's $2 tube of Colgate. It would be like progressive taxation, just without taxes or the government involved. But reality could flip this fantasy upside down. After all, who really has the time to comparison-shop? Is it the retired venture capitalist, or the single mom sprinting between two jobs and daycare pickup?

Time-starved, cash-strapped shoppers hit "Buy Now" precisely because they can't afford the time involved in the hunt. The algorithm reads their desperation in their browsing patterns—frantic evening sessions, no price-comparison tabs—and quietly nudges the prices upward.

Now add distress signals. Your phone pings from a downtown bar at 2 a.m.? That ride-hail "discount" you'd see at rush hour mysteriously vanishes—and if your battery's at 5%, the surge price might spike. Searching flights to Cleveland right after your high school friend posts an obituary? The algorithm picks up on your grief, quietly inflating fares, knowing full well you have bigger concerns right now than carefully comparing prices. No crystal ball required. Just location data, search history, and calendar access —permissions you clicked through two phones ago.

The Midnight Inhaler Mark-Up

Janelle's asthma flares at 1:42 a.m. Her smartwatch senses her elevated heart rate. Her phone, to which she gave microphone permission years ago, picks up her wheezing. She googles "urgent care near me," then clicks an ad for pharmacy delivery.

The pricing algorithm reads the tea leaves instantly. Middle-of-the-night shopping? Check. Search history including "urgent care" and "albuterol inhaler"? Check. Location data showing she's six miles from the nearest 24-hour pharmacy? Check. The price of a generic inhaler jumps from $24 to $39. The price of the vitamins stays at $12.99. Janelle hits "Buy Now," because what's fifteen extra bucks against the price of an ambulance ride?

For the drugstore, it's textbook economics: When you can't breathe, demand becomes perfectly inelastic (meaning you'll buy the inhaler regardless of its price). For society, it's price gouging with a silicon twist, personalized, invisible, and running 24/7. The algorithm doesn't sleep, and it probably knows exactly when you can't either.

When every breath, click, and heartbeat is data, stories like this aren't science fiction. They're business models. And when an executive makes a slip of the tongue, we get a rare glimpse of how companies really think about "personalized pricing."

Personalized Pricing Takes Flight

In 2025, *conspicuous* personalized pricing is a reality, and one of America's blue-chip airlines is bragging about it. Delta's president, Glen Hauenstein, stepped up to an Investor Day podium and more or less confessed the airline's new obsession: "We will have a price that's available on that flight, on that time, to you, the individual."[8] In other words, goodbye sticker price, hello bespoke fare. As of this writing, only three percent of Delta tickets are minted by machine-learning code. By New Year's Eve of 2026, the airline wants that share to quintuple to one in five.

How? Meet Fetcherr, a six-year-old Israeli startup that appears to be the algorithmic love child of a Wall Street analyst and an air-traffic controller. Fetcherr's AI pores over the digital bread-crumbs you shed when you browse—from frequent-flier number to favorite browser—and spits out a number it thinks you'll swallow. Hauenstein calls it a "super-analyst ... working 24 hours a day, seven days a week and trying to simulate ... real time, what should the price points be."[9]

If Delta's president can boast about algorithm-tailored airfares, just imagine how many other companies are quietly letting their own AIs size up your wallet without ever saying a word about it.

Fairness, Friction, and a Fork in the Road

Personalized pricing isn't automatically villainous. As far back as 1933 economists argued that differential pricing—a synonym for price discrimination—can actually improve consumer welfare overall, provided it boosts sales by enticing additional buyers with lower prices.[10] If algorithms primarily shower discounts on hesitant shoppers while charging regular customers no more than usual, then nobody loses and some gain. But something feels wrong. Fixed prices emerged from a moral conviction that all customers deserve equal treatment at the cash register, regardless of desperation or wealth.

Regulators face only bad options. If they ban personalized pricing entirely (as China did in 2022), then they also ban discounts otherwise offered to the needy. If they ban data collection, then they also cripple fraud detection and recommendation engines along with the pricing algorithms. Require disclosure?

Please—people already skip reading terms of service, which sometimes seem longer than *War and Peace*.

Even if regulators did try to ban differential pricing, enforcement might be impossible. The line between dynamic and personalized pricing blurs fast: If Delta hikes fares when demand spikes—and infers demand partly from who is browsing—did the airline adjust prices because of *when* you searched or *who* you are? How would regulators—or even a court—begin to answer that question?

When Sharpshooters Collide: The Competition Paradox

It might be tempting to think that personalized pricing is pure gravy for companies. But it isn't always. In a seminal model, Jacques-Francois Thisse and Xavier Vives (1988) showed that if two rival firms can both micro-target, they *might* end up in a prisoner's dilemma-type game.

In the classic prisoner's dilemma game, two suspects who've collaborated to commit a crime are arrested and separated by police. Each gets the same offer: Rat out your partner and go free while your partner does hard time. Stay silent and do moderate time. The cruel twist is this: If both snitch, both get screwed with long jail sentences. The rational choice for each individual (betray your partner) leads to the worst collective outcome (both rotting in jail).

Personalized pricing can work the same way. Each firm has a powerful incentive to adopt it, but once both do, they start undercutting each other on every individual customer. Research shows that this cutthroat undercutting on each individual customer might drive profits below the old uniform-price equilibrium.[11] The arms race helps shoppers and shrinks margins—exactly the opposite of what boardrooms envision. So the real winner in the personalization game might not be the companies deploying it, but rather the customers getting targeted in an example of what looks like corporate power backfiring on the corporation.

What is the takeaway? Algorithms can peek at your playlists or your panic and then decide, in milliseconds, whether you're a whale worth harpooning or a minnow worth baiting. Yet the same code that fattens one firm's quarterly earnings can ignite a profit-erasing war when competitors follow suit. Personalized pricing, like nitro-fuel, is powerful but volatile. Light it carefully, or the supposed profits blow up in your face.

Invisible Price Tags

If targeted prices seem sneaky—and they often do—wait until you discover the even subtler ways your wallet can shrink without you noticing. Algorithms don't always have to play with prices directly. Sometimes they quietly dismantle your options, eroding entire resale markets or slowly watering down your favorite content. As result, you end up paying the same, or more, for less. These changes happen so gradually, you might never realize what you've lost until it's long gone. We explore this idea in detail in the next part of this book.

Part Three
The Dispossession

10
Digital Serfdom
Is "Buying" Actually a Secret Rental?

In this chapter, the word *mine* (as in "belonging to me") takes on a different meaning. We're no longer talking about abstract notions of privacy. We're talking about real stuff, the things you buy and own, or at least think you do.

How Digitization Quietly Stripped Away Your Consumer Rights

What if I told you that the internet—the technology that promised to make everything cheaper—has actually made you poorer, not in obvious ways, but in hidden ways you've never noticed? Every time you click "Buy" on a website, you're unknowingly signing away rights your grandparents fought to secure. This chapter tells the story of how Silicon Valley pulled off the heist and why you didn't even notice it happening.

To trace this story, we have to start with a bit of history—and with one of those unlikely heroes who end up changing the rules of ownership, at least in the physical world.

WHO IS SUPAP KIRTSAENG? FROM STUDENT TO THE SUPREME COURT

Supap Kirtsaeng wasn't initially a crusader or an activist; he was simply a sharp-eyed graduate student at the University of Southern California who stumbled upon one of the most lucrative arbitrage opportunities available to a student. Kirtsaeng had discovered something peculiar: An international edition of a required textbook was selling in his home country, Thailand, for about $50. In the United States, students had to pay over $100 for the exact same book.[1] Same words, same pictures, and same knowledge—but double the price.

Recognizing an economic opportunity, Supap quickly enlisted his family in Bangkok to buy textbooks at local bookstores. They shipped these cheaper textbooks to his apartment in Los Angeles, where he sold them on eBay. Over two years, he imported about 500 different titles, generating roughly a million dollars in sales and netting profits of approximately $100,000, which helped to fund his education.[2] Not bad for a side hustle that required zero venture capital and no coding skills.

This entrepreneurial endeavor attracted unwanted attention. Supap's actions, known as "parallel importation" in the publishing industry, threatened to undermine publishers' ability to charge higher prices in wealthier countries. John Wiley & Sons, a billion-dollar heavyweight, struck back with a lawsuit, accusing him of copyright infringement. Think about that: A billion-dollar corporation sued a grad student for doing what every yard sale in America does daily—reselling legally purchased goods. John Wiley & Sons claimed the situation was different; textbooks targeted to foreign countries explicitly stated they could not be sold in North

America.[3] Was that restriction legal? The answer was unclear. Despite being outmatched legally and financially, Supap refused to back down.

The lawsuit exploded. A jury slapped Supap with $600,000 in damages in 2009, but he fought all the way to the Supreme Court.[4] By 2013, the justices faced a deceptively simple question: When you buy something, is it actually yours?

The case turned on the century-old first-sale doctrine, a legal principle stating that once you legally purchase an item—such as a book, CD, or artwork—you're generally free to resell, lend, or give it away without the original copyright holder getting involved or imposing extra conditions. Could publishers use geography to control what happens after the sale? Justice Breyer delivered the 6-3 verdict: Copyright law couldn't fence off resale markets by borders.[5] American consumers—who'd been paying the "rich country tax" while identical books sold at much lower prices overseas—finally caught a break.[6] Students' wallets at universities across the United States were suddenly thicker because one student decided to fight back. David had beaten goliath.

The victory party didn't last long. While Kirtsaeng was winning the right to resell physical textbooks, Silicon Valley was busy making physical textbooks, music, and video games extinct. The efforts of Kirtsaeng and countless other heroes responsible for our consumer rights were being undone by technological change, and almost nobody was asking the obvious question: Why should your rights as a consumer evaporate the moment a product goes from physical to digital?

Mario Brothers and the Vanishing Disc/Cartridge Economy

What do a Thai textbook smuggler's business and your dusty Nintendo cartridges have in common? They're both casualties of the same economic extinction event. Consider the old-school Super Mario Kart game disc or cartridge from your childhood. Back then, after spending months mastering each level, you could sell or swap your game with a friend or even a stranger at your local GameStop, getting back some cash to fund your next gaming adventure. This resale possibility wasn't just convenient; it also significantly influenced consumer behavior. Knowing you could recoup part of your investment later made consumers more willing to buy games in the first place, even expensive new releases. That $60 game was really a $35 game if you knew you could resell it for $25 later.

GameStop thrived precisely because of this economic reality. Its entire business model hinged on buying used games from consumers at low prices and reselling them at substantial markups. The numbers are staggering. In 2005, used games accounted for about a third of GameStop's total sales but generated well over half of its profits, according to the company's annual reports.[7] Put differently, GameStop made more money reselling one used copy of Madden than it did by selling two new copies. This lucrative model enabled GameStop stores to multiply rapidly across malls and shopping centers nationwide, shelves stacked high with pre-owned cartridges and discs.

Then came the asteroid that killed the dinosaurs—except this time, the asteroid was made of ones and zeros. The rise of digital downloads began chipping away at GameStop's seemingly invincible empire. As consumers increasingly preferred the convenience of buying games directly from digital storefronts or subscribing to online gaming services, GameStop's core business faced existential threats. The writing wasn't just on the wall. It was flashing neon, but GameStop kept rearranging deck chairs on the *Titanic*, focused on daily routines while catastrophe closed in.

Market forces soon revealed harsh truths. Between 2010 and 2020, GameStop's revenue dropped by nearly half, reflecting its struggle against the unstoppable tide of digital distribution.[8] Its stock price told an even grimmer story, dropping from $15 in 2007 to less than a dollar in 2020, and its once-celebrated business model became widely acknowledged as outdated and unsustainable.

Then, in 2021, things got weird. GameStop's share price skyrocketed, rising roughly 500-fold at its peak. What explained this amazing surge in GameStop's stock price? Retail investors rallied around GameStop shares as part of a David-versus-Goliath battle against major hedge funds betting against the company's survival through short selling. The irony was delicious: Wall Street had bet billions that GameStop would die, only to lose billions when an army of gamers decided to respawn it.

Despite this dramatic revival, no one seriously suggested that GameStop's fundamental business issues had vanished. Rather, investors targeted the company specifically because it was heavily shorted by professionals betting on its demise. GameStop became the world's costliest way to troll hedge funds.

ReDigi and the MP3 Resale Battle

As GameStop became resigned to bleeding out in shopping malls across America, a scrappy startup decided to fight. ReDigi had a simple, brilliant idea: Create "The World's First Online Marketplace for Used Digital Music." Think eBay for MP3s. ReDigi offered users the chance to legally resell digital songs purchased from platforms such as iTunes. The underlying technology was its Forensic Verification Engine, basically a digital bouncer that made sure you weren't trying to sell bootlegs. ReDigi's technology could verify you owned a song, delete it from your device, and transfer it to someone else—exactly like handing over a physical CD.

The music industry's response? It sued for $150,000 per song.[9] That's right—doing digitally what you could legally do physically could saddle you with a fine that cost more than a luxury car.

Music publishers relied on legal gymnastics. Laws protecting the resale of products explicitly covered physical goods, not digital ones. In other words, the law says you can sell atoms but not bits, even if the bits do exactly what the atoms did. Unless you're selling the original hard drive, digital resale remains unlawful. Imagine if Ford argued you couldn't resell your car because technically you'd be "copying" it from your driveway to someone else's.

ReDigi aimed to challenge this legal gray area, but the timing couldn't have been worse. Consumer interest rapidly shifted to streaming services like Spotify, where you pay $10 a month to rent millions of songs instead of $1 to own one song. Why fight for the right to resell something when a whole generation had already given up on owning it? Again, a consumer rights hero was undone by technological change. Though ReDigi's battle ended quietly, its story reveals the dirty secret of the digital age: The "Buy" button is lying to you. You're not buying anything. Instead, you're paying for an elaborate rental agreement with fewer rights. That arrangement might seem harmless, but the unseen consequences can be far-reaching.

THE MOST EXPENSIVE FREE BOOKS IN AMERICA

Libraries have long thrived on the same first-sale doctrine that protected Supap Kirtsaeng: Once a book is bought, the publisher cannot dictate how it's used, as long as it isn't copied illegally. This freedom allows libraries to buy one book at the same price a consumer pays, then lend it out until it starts coming apart at the seams. A single copy of Harry Potter might pass through 100 different hands over a few years before falling apart, and J.K. Rowling can't do anything about it.[10] The same principle powered the business model of video rental giants such as Blockbuster. They bought movies at retail prices and rented them countless times, turning a $20 DVD into $2,000 in rental revenue —perfectly legal, wildly profitable.

Libraries are America's underappreciated economic equalizer. They offer free access to books, enabling children and adults alike to read, learn, and explore new worlds and ideas without the burden of direct cost. A kid in the Bronx gets the same access to knowledge as a kid in Beverly Hills. That's the American Dream with a library card.

But publishers found their loophole. Because the first-sale doctrine doesn't cover digital books, they sell libraries expiring e-book licenses—often $60 each, lasting only 26 loans or two years, whichever occurs first. Meanwhile, consumers can "buy" the same title on Amazon for $15 to keep, but never share.[11] Publishers implement these constraints explicitly to encourage readers to buy their own copies, essentially charging libraries a 300% tax. Once books go completely digital, publishers' leverage skyrockets, and so will the price quoted to libraries for copies that can be lent out.

Let's envision the dystopian scenario. Imagine publishers entirely halt the printing of physical books, shifting completely to digital distribution. Suddenly, every library in America becomes a subscription service that can't afford its own subscriptions. Without the legal protections that supported endless lending, libraries will find themselves powerless, unable to provide open and free access to books.

Now consider the real cost: Knowledge becomes a luxury good, even more so than is currently the case. Your zip code already determines your vocabulary to some extent, and your parents' paycheck predicts your intellectual achievements. Nonetheless, a motivated disadvantaged student with a library card still has nearly endless knowledge at their fingertips. If we gut libraries, that access vanishes. Publishers are architecting a caste system where literacy itself becomes hereditary—unless more books become nearly free, as Chapter 14 suggests they might.* Then the whole world becomes a library, and the gates come down.

Your Rights, Digitized Away

Digitization reshapes far more than just the way we consume media. It silently dismantles consumer rights that took over a century to build. Ironically, we lost these rights at the exact moment they would have become infinitely valuable. Physical resale was nice. Digital resale would have been revolutionary. We'll never know what we missed.

Think about the underlying economics. With physical products such as cars, waiting to buy a used version inherently means settling for lower quality—scratched paint, worn seats, mysterious smells. But digital products, unlike physical goods, do not degrade. A "used" MP3 is molecularly identical to a "new" MP3. There's literally no difference except the price. Buying a used MP3 offers consumers the same quality as a new MP3, but at a lower price.

Firms saw this profit catastrophe coming and killed it in the crib. By strictly controlling digital licenses and usage rights, companies have eliminated the concept of a "used" digital product entirely, creating a world where every customer is a first-time customer, forever. This loss of resale markets—a staple of economic freedom for decades—means that consumers pay full price in perpetuity for products that cost virtually nothing to reproduce.

Higher profits for publishers and creators offer a hidden silver

* Platforms set minimum prices for e-books, and I am happy to play by their rules. They've built great platforms, and I am glad to do my part to keep them thriving. Someday, these rules may change.

lining: more incentive to produce new, high-quality content. This is the classic producer's defense: "Sure, I'm overcharging you, but look at all this great new product I can afford to make!" With increased revenue potential from every sale, creators and companies are motivated to invest in innovative and improved digital products.

We're left choosing between two imperfect extremes. Allow publishers to keep restricting how digital copies can be reused, and we get a world where profits rise at consumers' expense, especially for those least able to pay. But extend full lending and resale rights to digital files, and a single copy could be shared infinitely, eroding sales and the incentive to create new work. For now, society has chosen the first path: We've traded ownership for a vast catalog of things we can buy but never sell or truly possess.

The real tragedy may be psychological. When you next see that dusty Mario Brothers cartridge lying forgotten in your attic, you're looking at the last generation of products you actually owned. Digitization brings convenience, but at a hidden cost, turning purchases into permissions and buyers into borrowers.

This change is largely lost on today's youth, who grow up in an environment mostly devoid of used games, resold textbooks, and other second-hand media. Ask anyone under 25 if they've ever sold a used video game, and they'll look at you like you asked if they've ever sent a telegram. To them, you don't buy things—you buy access to things. The idea of "owning" a movie seems as antiquated as using a horse for transportation.

We haven't just lost our rights—we've lost the memory that we ever had them. In one generation, we went from owners to digital serfs, and we did it willingly because the user interface was pretty.

And yet, even where resale markets still survive, technology still finds ways to tilt the game. The next chapter begins by exploring how e-commerce is reshaping resale markets for physical goods—and if you think you already know how, you're probably wrong. Don't feel bad, though. The economics profession got it wrong too, at least for a while.

11
The Savvy Consumer Delusion
Can You Beat the System?

The Inverted Bazaar

The history of technology is basically a highlight reel of humanity creating technology to fix problems but inadvertently creating new problems we never saw coming. Email freed us from phone tag, then enslaved us to our email inbox. Social media connected us to long-lost friends, then made us hate some relatives. Smartphones put the world's knowledge in our pockets, then turned us into hunched-over zombies who walk into traffic.

My favorite examples of technological backfire come from the digital economy, where our attempts to outsmart the system keep outsmarting us instead. I'm not talking about bugs or glitches or evil corporate conspiracies (mostly). Instead, I'm talking about the natural result of millions of rational people making sensible decisions that, taken together, create perverse, even bizarre, outcomes.

This chapter analyzes two digital advances with surprising consequences: online resale markets displacing local sales and ad-blocking technologies reshaping the web. Each innovation solved genuine problems. Each also triggered cascading effects worth examining closely.

Consider online marketplaces for used goods. They were supposed to create perfect competition—millions of sellers competing on price, driving costs down to theoretical minimums. For new products, that's exactly what happened. Amazon crushed retail margins like a steamroller powered by pure capitalism. In contrast, something weird happened with used goods. Even when the law protects your right to resell, you're still getting fleeced. To understand why, read on.

The Great Online Paradox of Used Goods

Two MIT economists discovered a phenomenon that should have broken the laws of economics: Used books cost MORE online than in stores, despite more competing sellers and lower operating costs.

Here's the beautiful stupidity of physical used bookstores: Imagine walking into a local used bookstore looking for a very specific and unusual title—say, *Frogs into Princes: Neuro Linguistic Programming* by Richard Bandler and John Grinder. You scan the shelves carefully, hopeful but realistic. The store has a less than 0.01% chance of having this hard-to-find book. And, indeed, the store doesn't have it. Your desired book is surely out there, somewhere, but maybe it's hundreds or even thousands of miles away. Frustrated, you settle for another title, something interesting but not precisely what you wanted, and you haggle the price down to $3 because, hey, it's not even what you came for.

Then the internet flipped the script. Suddenly, location became irrelevant. Instead of settling for second-best, you can find that exact NLP book from a seller in Tucson who's been waiting three years for a like-minded person to buy it. Intuitively, this precise matching might seem like it would lead to lower prices—after all, more sellers usually mean more competition. But Glenn Ellison and Sara Fisher Ellison discovered something surprising: Prices for used books online often exceed those found in bricks-and-mortar stores.[1]

What's happening here? Online platforms don't just offer more books—they also create perfect marriages between obscure books and the obsessives who desperately want them. Unlike a random bookstore encounter, they turn every transaction into a micro-monopoly. You're not buying "a book." You're buying *the* book, and the seller knows it. According to the Ellisons' findings, online markets do lower costs and boost competition, but this enhanced matching effect ultimately dominates, leading to higher average prices overall.

The outcome? Consumers win and lose simultaneously. Consumers pay more, but they get exactly what they want. Still, it is strange that online marketplaces are raising the prices of some used products. Of course, this paradox will become moot if publishers can kill the used market entirely. You can't have expensive *used* e-books when *used* e-books can't exist.

At least we'll still have the "free" internet, funded by ads. Probably. Maybe?

THE GREAT AD-BLOCKING PARADOX: HOW TRYING TO IMPROVE YOUR INTERNET EXPERIENCE MIGHT ACTUALLY BE RUINING IT

Picture this: You're trying to read an article about, say, the mating habits of seahorses. But before you can learn about how male seahorses carry the babies, you're assaulted by the following:

- A full-screen video ad for car insurance that auto-plays at maximum volume
- Three banner ads for products you Googled once six months ago
- A pop-up begging you to subscribe to a newsletter
- Another pop-up asking if you'd like to enable notifications
- And somehow, inexplicably, an ad for adult diapers (the internet's algorithms have concerns about your future)

By the time you've closed, muted, and X'd out of this digital obstacle course, you've forgotten why you even cared about seahorse reproduction in the first place.

Enter the ad blocker: a browser extension that promises to vanquish these digital distractions. Click to install it, and *poof*—the internet becomes a serene, ad-free paradise. It's like noise-canceling headphones for your eyeballs.

If you're one of the roughly 25% of internet users who've installed an ad blocker, you probably felt like you'd discovered fire. Websites load faster, articles become readable again, and YouTube videos start instantly without forcing you to endure a 30-second insurance ad. What could possibly go wrong?

As it turns out, quite a lot. And I say this as someone who despises online ads with the burning passion of a thousand suns.

Digital Bouncers at the Internet's Worst Party

Before we dive into the economic apocalypse, let's understand the elegant simplicity of ad-blocking technology. When you visit a website, your browser sends out requests to various servers. Some of them deliver the content you want (seahorse facts), others deliver ads (adult diapers). Ad blockers maintain a bouncer's list of known ad servers. When your browser tries to contact "annoying-ads-r-us.com," the ad blocker steps in like a protective parent: "We don't talk to strangers."

There's no shortage of ad blockers. The best known, Adblock Plus, was created in Germany in 2006 by Wladimir Palant, presumably after one too many pop-up ads ruined his morning coffee. A decade later, more than 500 million downloads proved he wasn't the only one fed up.

The Beautiful Economic Theory That Went Horribly Wrong

Here's what economists predict should (initially) happen with ad blocking. Ads are essentially a "price" you pay to view content—not in dollars, but in annoyance, time, and cognitive burden. When you block ads, you've effectively made websites "cheaper" to visit. Following basic economic logic, when prices drop, consumption should increase. People should spend MORE time online, visit MORE websites, and consume MORE content.

Initially, that's exactly what happened. Joel Waldfogel, Johnny Ryan, and I studied thousands of websites and found that in the short term, sites with ad-blocking visitors saw increased traffic. It's like a store having a sale—customers flood in when the "price" drops. But then something unexpected happened. And by "unexpected," I mean "completely predictable if you think about it."

WHAT IS THE WEB BROWSERS' DILEMMA?

Remember the prisoner's dilemma described in Chapter 9? With ad blockers, we've created a digital version of the prisoner's dilemma, and we're all simultaneously the criminals, the police, and the prison. Every time you install an ad blocker, you're choosing the "defect" option—that is, making yourself better off while making the system a little bit worse for everyone else by depriving websites of the revenue needed to invest in new, high-quality content. Just like those two prisoners in the dilemma, when we all make the smart individual choice of blocking ads, we all end up worse off than if we'd cooperated by tolerating those annoying ads.

Here's a concise timeline of how ad blocking backfired:

Phase 1: The Honeymoon (2013–2014). Early adopters install ad blockers. Websites still have enough paying customers (those seeing ads) to maintain quality, or they don't respond to shorter-term drops in revenue by firing people. Ad-blocking users are essentially freeloading off the ad-viewing masses, like sneaking into a concert while others pay for tickets. Life is good.

Phase 2: The Tipping Point (2015–2016). Ad blocker usage explodes from a niche tool to mainstream adoption. Suddenly, 20-30% of visitors at some sites aren't generating revenue. Websites start noticing their bank accounts looking anemic. They start thinking about resizing their employee count.

Phase 3: The Death Spiral (2016–Present). Sites with heavy ad-blocking usage don't just lose revenue—they start producing less content. Word counts drop. Ads increase.[2] And, even with "free" (ad-blocked) access, users start visiting these degraded sites LESS.

The numbers are staggering. For every 1% increase in ad-blocking visitors, sites saw a 0.67% decrease in traffic over time. Sites with 20% of users blocking ads saw traffic drop by 13%. The revenue impact was even worse. That same 20% ad-blocking rate could mean a 30% revenue loss, because the website is losing ad revenue not only from users who install ad blockers but also from internet users without ad blockers who abandon the now-crappier site.

The shift toward mobile browsing has given publishers a temporary reprieve: Ad blocking is still about two and a half times more common on desktop than on mobile.[3] That imbalance has helped steady revenues in the short run, though it also means that today's web likely offers slightly less content—and of lower quality—than it would in a world without ad blockers. Yet the deeper threat remains, with consequences that reach well beyond media economics into the realm of inequality.[4]

Do Ad Blockers Worsen Inequality?

Here's an idea that might blow your mind: Advertising is secretly one of the most redistributive economic systems we've stumbled into. Stay with me here.

Yes, everyone sees ads, but not everyone is equally valuable to advertisers. Wealthier users, with greater disposable incomes, attract higher advertising bids, generating more revenue for the website per visit. Other users, especially those with lower incomes, see the same number of ads but bring in far less revenue. The result is a hidden form of redistribution: Wealthier visitors essentially subsidize the experience for everyone else.

This subtle redistribution vanishes once websites move from ad-supported to subscription-based models. Subscriptions typically involve a flat fee, priced to attract affluent users who willingly pay and inadvertently pricing out lower-income visitors who can't or won't pay for a subscription. The very users who previously benefited from the indirect subsidy of advertising now lose access entirely.

In short, advertising quietly enables a surprisingly equitable outcome, letting everyone participate, even those whom advertisers consider less valuable. Ironically, it's the spending habits of the wealthy, eagerly targeted by advertisers, that fund open access for everyone else. Add blocking destroys this happy equilibrium. Think about who installs ad blockers. They tend to be:

- Tech-savvy (obviously)
- Higher income (they put a premium on their time)
- Extremely irritated by interruptions (zero patience for anything slowing down their Amazon shopping spree)

In other words, ad blockers are often adopted by precisely the users that advertisers most desperately want to reach. These people are less price-sensitive, flush with disposable income, and more likely to impulse-buy things like a $200 smart water bottle that politely reminds them to hydrate every 90 minutes. Plus, because they're actively dodging ads, they're far less likely to know about your competitor's products, making them even more valuable as customers. When these prime eyeballs disappear behind digital fortresses of ad blockers, advertisers naturally pull back their budgets. It's like a high-end cocktail bar suddenly losing all its customers who order $18 gin concoctions infused with artisanal rose petals. The bar stays open for now, but it's left serving $3 beers to folks who nurse a single pint all night.

Without those prime eyeballs, the website's financial model starts to unravel. Writers get laid off, content quality nosedives, and suddenly the whole experience deteriorates for everyone. In short, ad-blocking isn't merely about personal convenience; it's also quietly rewriting the economic rules of the entire internet, and not in a good way.

There Are No Good Answers, Only Different Flavors of Bad

I wish I had a solution that doesn't stink. But every option has massive downsides.

Option 1: Block the ad blockers. You might think that websites can force people not to use ad blockers, but this option doesn't work in reality. Bild.de, Germany's most popular tabloid website—think of it as the digital equivalent of a neighbor who gossips loudly over the fence, sometimes sharing content not suitable for minors—decided in 2015 to do something about ad-blocking freeloaders. It went full nuclear: Stop blocking ads, or you can't read our articles. What happened next was like watching a bar implement a strict dress code. Bild.de's traffic rank decreased by 10% in just three months as ad-blocking users fled to competitors. Forbes tried a softer approach, politely asking ad-block users to disable their blockers. The result: 58% of visitors basically said "Nah" and left to get their business news elsewhere. It turns out people really, really hate ads. Who knew?

Option 2: Charge a subscription fee. "Just charge a subscription fee!" you might suggest, perhaps while adjusting your monocle. How will that strategy play out? The New York Times successfully charges for digital subscriptions. But they're the NEW YORK TIMES. What about "Jim's Seahorse Facts Blog" or "Decent Recipes That Don't Require Saffron"? Would you pay $5 a month to visit these websites?

More importantly, consider the drive-by visitor. Maybe you need Jim's seahorse expertise exactly once, for a crucial trivia night question. Under an advertising model, Jim still gets a few cents from your visit, and those pennies from many drive-by visitors can eventually add up to substantial cash. But under a subscription model, Jim gets very little from potential drive-by visitors, and you're stuck with inferior seahorse knowledge.

This is why even successful subscription sites like the New York Times site maintain some free, ad-supported content. They need to capture value from casual visitors while converting regular readers to subscribers.

Option 3: Get everyone to disable ad blockers (willingly). Collective action problems don't solve themselves through wishful thinking. They follow the same logic as the "web-browsers' dilemma" described earlier. Specifically, expecting individuals to act against their own short-term interests for the public good rarely works.

Option 4: Offer government subsidies. But do we really want the government to fund journalism? Typically, that leads to either bland content or state propaganda.

Option 5: Accept the death of the open web. Maybe the internet becomes like cable TV—packaged subscriptions for bundles of sites. Hope you like the "Lifestyle Bundle," because that's the only way to get cooking sites AND fitness advice!

The catch? Prices rise quickly. Just ask anyone paying $300+ a month for cable and streaming services that focus on entertainment (for example, Netflix, Hulu). Now imagine adding bundles for news, research, hobbies, health information, and everything else you casually browse today for free. Budget-conscious users who can't justify the expense get locked out entirely, losing access to the resources that used to be freely available to anyone with an internet connection. The open web becomes a gated community for those who can afford the cover charge.

Option 6: Merge content and ads into one indistinguishable mess. Maybe ads slots just disappear from the internet—not because they're blocked, but because they've evolved to be invisible. Imagine a scenario where your favorite cooking blog subtly praises a certain olive oil—not because it's objectively superior, but because the olive oil company quietly sponsored the article. Or perhaps that travel review you trusted was funded by the city's tourism board. The line between authentic content and hidden advertisements blurs until you can't tell one from the other. Suddenly, the entire web becomes a series of sneaky infomercials disguised as unbiased journalism, recipes, or life hacks. Welcome to an internet where everything is "content," and ads are hiding in plain sight. Advertiser-supported content and product placement are already routine on YouTube and TV.[5] When it spreads to esteemed news organizations, the question won't be what's true—it'll be who paid for it.

The Lesson Hidden in the Rubble

The ad-blocking paradox teaches us something profound about unintended consequences in the digital age. We thought we were just avoiding some annoying ads. Instead, we might have accidentally destroyed the economic model that democratized information for billions of people. It's a reminder that in complex systems, obvious solutions often backfire spectacularly. Every time we find a "simple" fix—block ads, connect companies with their ideal consumers, disrupt an industry—we discover new and exciting ways for things to go wrong.

As I write this, I'm using an ad blocker. Yes, I'm part of the problem. But I challenge you to browse the modern web without one and maintain your sanity. An ad blocker is the rational choice, and maybe that's the real lesson: Sometimes, knowing better isn't enough to do better. We're all trapped in a prisoner's dilemma where the rational individual choice (block ads) leads to an irrational collective outcome (worse internet for everyone).

Websites Stop Caring About Ad Blocking Because They Won't Exist

Ad blockers have chipped away at publishers' revenue, but generative AI could erase their business altogether. Cloudflare, a content delivery network that helps speed up and secure about 20% of the internet, recently analyzed how users get information online. Under the old implicit deal, you gave Google your website's content and it sent traffic in return. Swap in ChatGPT and the chance of getting a click-through to other websites is 750 times lower than it used to be. Ask Anthropic's Claude and click-through rates fall 30,000-fold.[6]

No website, whether it's Jim's Seahorse Facts Blog or a major news outlet, can survive such a drastic reduction in visitors. Without traffic, all sources of revenue dry up: no ads, no subscriptions, no financial incentive to keep publishing content. It's the digital equivalent of running a store with no customers walking through the door. If AI platforms continue to pull content from websites without sending visitors to those websites, publishers won't just lose revenue and degrade content—they and their websites will cease to exist. The open web as we know it could disappear, replaced by AI-generated summaries of the content it once hosted.

People may look back with disbelief on the era when tens of millions of unique websites flourished. To future generations accustomed to receiving their information exclusively through a few AI gatekeepers—ChatGPT, Google's Gemini, Claude, and others that haven't launched yet—it may seem almost unimaginable that the internet once offered almost endless diversity rather than curated summaries from a tiny handful of platforms.

At that point, who will create the fresh content that these AI agents summarize? We might find ourselves with a seemingly all-knowing (but secretly fallible) assistant—at least, about anything published before 2026.

The Moral of the Story (If There Is One)

Ad blockers, generative AI, and used books websites have at least one thing in common: In all three cases, technology that seemed to obviously benefit consumers (locating the rare book you want, getting instant answers from AI, blocking annoying ads) creates unexpected economic consequences that makes things more expensive or worse for everyone.

It's almost like the universe has a twisted sense of humor about market efficiency. Or maybe it's just that every time we think we've outsmarted the system, the system outsmarts us right back. Either way, I'm keeping my ad blocker, using generative AI instead of web searches, and paying too much for used books. Some principles are worth being a hypocrite for.

Call it my tiny act of rebellion. Each small step feels like a way to push back against the business juggernauts perfecting the art of making us spend more than we ever planned to. Ironically, though, consumers (and I) aren't suddenly gaining an edge over firms. The same clever code that helps us block ads also helps them study us—and steer us—straight toward choices that drain our wallets. Read on and learn to save money twice—once by learning the digital playbook and seeing through the illusion, and again by being too absorbed in this book to fall for it.

12

Confuse & Conquer?

How Sellers Hacked the Low-Price Economy

The last chapter examined innovations that initially appeared to benefit consumers. This chapter examines innovations with the opposite goal: tactics explicitly designed to extract your money. We'll explore practices that skirt a fine line, remaining legal (for now) but ethically questionable. Then we'll examine the technologies consumers can deploy to fight back. In doing so, we'll reveal a surprising twist: Some of the tech behemoths we love to vilify, such as Amazon, might actually be our allies in this particular battle.

Does Your "Smart" Phone Make You a Dumber Shopper?

Jamie Chen considered herself financially savvy. The 34-year-old, Seattle-based marketing manager tracked her spending, maxed out her 401(k) contributions, and even kept a color-coded spreadsheet of her monthly utility expenses. Yet when she audited her credit card statements in January 2025, she discovered something shocking: She had spent $987 the previous year on subscriptions she never used, and another $147 on sneaky fees. There was the meditation app she'd downloaded during a particularly stressful week in March (used twice, $11.99/month ever since). The premium LinkedIn account from a job search two years earlier ($29.99/month). A "free" trial for a meal-planning service that auto-converted to a paid subscription ($9.99/week). And buried in her Ryanair booking from her summer vacation, a $47 seat selection fee that appeared only after she'd entered her credit card information.

Jamie isn't alone. According to a 2025 CNET survey, the average American wastes $204 annually on unused subscriptions. For Gen Z, that number jumps to $274.[1] Here's the truly puzzling part: Jamie, like most of us, carries a device in her pocket that can access virtually all human knowledge, compare prices across thousands of vendors quickly, and even remind her to cancel subscriptions. So why is she—why are we—getting fleeced more than ever?

The answer reveals something profound about how the digital economy really works. Although technology was supposed to create perfectly efficient markets—economist-speak for a world where consumers always get the best deal—it has instead spawned an arms race of manipulation. Companies now deploy artificial intelligence, behavioral psychology, and big data not to serve us better, but to extract maximum profit from our psychological weaknesses. Welcome to the dark side of digital commerce, where your smartphone doesn't make you a smarter shopper—it makes you a more profitable target.

THE SEARCH FRICTION PARADOX

In 1999, as internet firms were taking off, economists made a bold prediction: The internet would spell the death of price markups. The logic seemed unassailable. In the physical world, comparing prices required driving from store to store, making phone calls, or poring over newspaper circulars. Online, though, a few clicks would reveal every option, forcing sellers to compete on price alone. The *law of one price*—the economic principle that identical goods should sell for identical prices—would finally reign supreme.

Fast forward to today, and economists Glenn Ellison and Sara Fisher Ellison must be having a good laugh. Their groundbreaking study, "Search, Obfuscation, and Price Elasticities on the Internet," revealed that far from eliminating price dispersion, the internet had simply forced sellers to become more creative.[2] The Ellisons studied Pricewatch, a computer parts comparison site, and discovered something remarkable: Firms were deliberately creating confusion to maintain their markups.

Here's how it worked. Pricewatch ranked sellers by lowest price first. Naturally, every seller wanted to be listed first. But rather than engage in a race to the bottom, clever retailers found a workaround. For a rock-bottom price, they'd offer a base product that was essentially worthless on its own—say, a computer memory module for $19.99. This "bait" product would win the company the top ranking. But when customers clicked through, they'd discover that the memory module needed a heat sink to prevent overheating and failure. The "complete" options with the heat sink pre-installed cost an additional $15, a substantial markup that far exceeded the actual cost of the component. Other sellers used different strategies to stand out. One might not even sell the useless base product—forgoing the top ranked slot—but offer the combined product at a more reasonable price, banking on developing a reputation for reasonable prices. In the end, prices remain dispersed across sellers.

Firms are essentially solving an optimization problem. The question is how they can win the price comparison while still maintaining profit margins. The answer is to make true comparison impossible, and this obfuscation has become ever more sophisticated. Consider some of today's tactics:

Product proliferation. Walk into any pharmacy and try to buy Tylenol. You'll see Extra Strength Tylenol, Tylenol PM, Tylenol Cold & Flu, Tylenol Arthritis Pain—over 30 variants in total. Each has slightly different formulations, dosages, and prices. Quick: Which offers the best value per milligram of acetaminophen? This complexity isn't a flaw in the company's product strategy—it's the whole point. You think you're comparing apples, oranges, and grapefruits, but you're really just staring at slightly different shades of apple.

Retailer-exclusive model numbers. That 55-inch Samsung TV at Best Buy isn't technically the same as the 55-inch Samsung at Amazon, even though they look identical. The model numbers differ by a single digit—UN55TU7000FXZA versus UN55-TU7000BXZA—making price comparison difficult. A producer could design dozens of nearly identical TVs exclusively for different retailers for the main purpose of making comparison shopping harder.

Drip pricing. The advertised price is just the beginning. Airlines pioneered this art, but it's spread everywhere. StubHub shows you a concert ticket for $75, but at checkout, you're paying $97.50 after service fees, processing fees, and delivery fees (for an electronic ticket!). Recent research by the UK's Competition and Markets Authority found that drip pricing causes consumers to pay 15-20% more than they would if all fees were disclosed upfront.

Why does drip pricing work? After spending ten minutes

choosing seats, entering your details, and creating an account, abandoning your purchase over an extra $22.50 feels like admitting defeat. At work here is what economists call the *sunk cost fallacy*: Once you've invested time and effort, you're irrationally inclined to keep going. If you were operating rationally, you would ignore the effort already spent and ask yourself: "Do I still want this ticket at $97.50?" If the full price had been clear upfront, you might never have chosen to purchase the ticket in the first place.

Drip pricing harms you even if you abandon the purchase after seeing the inflated final price. All the time you spent entering your information—only to back out at the last moment—is completely wasted. Had the true cost been displayed upfront, you'd have skipped the whole ordeal.

Ryanair provides a particularly egregious example. The airline advertises flights from London to Barcelona for £14.99. Sounds great! But that's just the beginning. Want to bring a carry-on bag? £27. Have a stroller? £15. Select a seat? £2-15, depending on location. Fast track at the airport? £8. Insurance? £13. By the time you board, that £14.99 flight costs £90.

Ryanair's CEO Michael O'Leary even floated the idea of charging passengers to use airplane bathrooms. Imagine having to swipe your credit card at 30,000 feet just to answer nature's call.[3] While the bathroom fees never materialized, the idea perfectly captures the drip-pricing mentality: Give away the core product, then charge for everything that makes it usable. O'Leary even once said, "I have this vision that in the next five to 10 years that the air fares on Ryanair will be free."[4] He wasn't joking. Of course, free before the standard extras isn't the same as truly free.

The Subscription Trap

Perhaps nowhere is digital-age consumer manipulation more evident than in the explosion of subscription services. Your razor

(Dollar Shave Club). Your music (Spotify). Your video games (Steam). Your vegetables (Imperfect Foods). Your socks (Bombas). Your air filters (FilterEasy). Even common household items (Amazon Subscribe and Save).

The business model is brilliant in its simplicity: Leverage human psychology to create revenue streams that flow long after the customer stops receiving value. Here's how it works.

Step 1: The Irresistible Offer. "Start your free 7-day trial!" The word "free" triggers what behavioral economists call the *zero price effect*. Our brains essentially short-circuit when something costs nothing, dramatically overvaluing the offer. Dan Ariely's famous experiment at MIT generated evidence in support of the zero price effect: When chocolates were priced at 15 cents versus 1 cent, 73% chose the more expensive option. But when the price of each dropped one penny, implying the tradeoff was now 14 cents versus free, suddenly 69% wanted the formerly 1-cent chocolate.[5]

Step 2: The Friction Asymmetry. Signing up takes 30 seconds. Three clicks, enter your credit card (you know, just in case you want to continue after the trial), and you're done. But canceling is a different story. First, you can't find the cancel button—it's buried three menus deep, if it's there at all. When you finally locate it, you're required to call the company during business hours. And then you have to wait to speak to someone, who tries to convince you to maintain your subscription. The average American consumer spends 15 minutes a week on hold.[6] When you finally reach someone, they offer you a "special retention offer" if you keep your subscription for now. If you want to cancel later, you'll need to start the entire process from scratch.

Step 3: Status Quo Bias. Even when canceling is easy, most people don't. Behavioral economists call our powerful tendency to stick with default options the *status quo bias*. The classic demonstration comes from a study of employee retirement plans. When workers had to actively opt into their company's 401(k) plan—even with generous employer matching—only 70% bothered to enroll after two years on the job. The other 30% were literally leaving free money on the table. But when the company made enrollment automatic with the option to opt out, 95% of employees remained enrolled after two years.[7] Same benefit, same choice, radically different outcomes. The only difference was the default setting. Status quo bias helps explain why subscriptions are so lucrative: Once you're in, inertia keeps you paying.

The gym industry perfected this technique long before Silicon Valley caught on. The average gym member visits 4.3 times per month. With a monthly fee of $70, the per-visit cost is about $17—or $7 more than a day pass that costs $10.[8] But gyms know that your optimistic January self ("This is the year I get in shape!") will sign up for an annual membership, while your realistic February self is too embarrassed to admit defeat by canceling.

Digital subscriptions took this model and turbocharged it. They have perfect information about usage, they can personalize retention tactics, and they face no physical constraints on the number of "members" they can accommodate. A gym might feel guilty about having 10,000 members but only 3,000 capacity. But Spotify? It's happy to have millions of subscribers who haven't opened the app in months.

THE AI ARMS RACE: PERSONALIZED PREDATION

Obfuscation and subscriptions have worked reasonably well for companies, but they're no match for the precision of personalized tactics powered by artificial intelligence. Companies now build detailed psychological profiles of every customer, allowing them to customize their tactics to customers' individual weaknesses.

Consider sports betting companies. Their AI presumably doesn't just track the bets you make. It can also build a comprehensive model of your psychology. Do you chase losses? Or do you pause use? These companies can design their app to offer you "boost tokens" after a bad beat. Do you bet more when your team is mentioned? If so, you can expect push notifications minutes before every home game. Are you likely to quit after losing a certain amount? Here comes a perfectly timed "free bet" to keep you engaged.

These tactics have forever changed the lives of people like Jordan Holt, who tried to quit gambling but was drawn back in. Jordan bet a whopping $878,529 in 2023. He tried to quit, but Fan Duel gave him $7,839 in enticements over the year to continually lure him back.[9]

In truth, retention offers and other behavioral manipulation tactics have long been used. But just like personalized pricing, they now can be targeted with far more precision. Maybe their historical data suggest that you need a $100 bonus to remain a customer, but I get offered only $25. These two different offers are another form of personalized pricing.

The Counter-Revolution

Digital manipulation is so pervasive that fighting back has become big business itself. Consider Truebill, an app built specifically to protect consumers from hidden subscriptions and sneaky charges. Its reward was a staggering $1.275 billion acquisition by Rocket Companies.

Truebill is not alone in offering valuable consumer services. Trim negotiates bills on your behalf. Honey automatically applies coupon codes. Bobby tracks all your subscriptions in one place. When protecting people from manipulation becomes a billion-dollar industry, you know manipulation has gone mainstream. The result is a bizarre arms race. Rocket Money's business model depends on subscription services being predatory—if canceling were easy, who would pay $12/month for an app to do it for you? (Yes, Rocket Money is itself a subscription. The irony is not lost.)

The numbers tell the story. By the time Truebill was acquired in 2021, it had reached 1 million customers and saved them over $100 million in unwanted subscriptions.[10] Coincidentally, its anticipated revenues from these customers were also $100 million. The predatory subscription model is so profitable that there's room for an entire industry devoted to fighting it, and that industry is itself quite profitable.

Looking forward, AI-powered tools may shift the balance further. Imagine uploading that 47-page terms of service agreement—the one written by lawyers hoping you'll never read past paragraph three—and getting back a plain-English summary highlighting every trap door and gotcha. Chat-based AI tools can decode the legalese, benchmark competitors' offers, and help users ask the exact questions companies were hoping they wouldn't think of. Running a subscription agreement through ChatGPT before signing could become as automatic as checking reviews on Amazon.

Some consumers are going even further and exploiting the loopholes in firm's marketing strategies. "Subscription hopping" has become a sport among cash-strapped millennials—signing up for free trials with virtual credit card numbers that auto-expire, using different email addresses to claim multiple trials, and maintaining elaborate spreadsheets to track which services they're currently scamming. Reddit's r/frugal has over 6 million members sharing tips on gaming the system.

None of this should be necessary. The fact that consumers need to employ subterfuge to avoid being overcharged reveals how broken the system has become. Yes, personal responsibility matters, but markets shouldn't be designed to make responsible budgeting so difficult.

The Platform Solution

Platforms themselves might be part of the solution, depending on how they structure their marketplaces. For example, think about platforms such as eBay and Amazon. While Amazon does sell some products directly and might want to raise prices on those items, it also hosts millions of third-party sellers. In fact, third-party sellers account for about a third of the sales on Amazon's website.[11] This creates an interesting incentive: Rather than maximizing profits from individual products, Amazon's primary goal may be to keep consumers coming back to the platform so that Amazon continues generating revenue from each third-party transaction. If third-party sellers' profits get squeezed, that doesn't hurt Amazon's bottom line, at least not directly. This dynamic is even more pronounced for eBay, which operates almost exclusively as a marketplace without selling items directly.

In essence, these platforms don't want to use the deceptive practices that extract more money from consumers but make them less likely to return. Instead, the platforms have strong incentives to facilitate good prices and positive shopping experiences.

A fascinating study by economists at Stanford University and the University of Chicago demonstrated how platforms can achieve these goals simply by changing their search algorithms.[12] At the time of the study's onset, eBay's search algorithm worked much like Craigslist. When consumers searched for an item, they saw auctions or "Buy It Now" listings that matched their search terms, sorted according to a relevance algorithm that mostly focused on the date the item was posted or the date the auction was ending. Under this sorting method, it was very difficult for consumers to find which sellers offered the best prices. They could do it, but it required extensive searching and comparison.

Amazon, by contrast, had developed a more consumer-friendly approach. It first sorted results by specific product categories. When consumers clicked on a product, they saw a dedicated product page with a Buy Box featuring a recommended seller, as well as a list of other sellers offering the same item. Crucially, the Buy Box wasn't random. Instead, Amazon had designed the algorithm to feature sellers who combined low prices with strong reputations and high ratings. Essentially, they wanted customers to have positive experiences so they would return to the platform.

The eBay study investigated what would happen if eBay adopted Amazon's strategy. The researchers conducted both controlled experiments and simulations, and they found three benefits from the new sorting approach. First, the new sorting algorithm helped consumers identify the particular sellers who offered better prices. Second, consumers spent less time searching, as evidenced by fewer clicks per browsing session before making a purchase. Third, and perhaps most importantly, all sellers now had stronger incentives to lower their prices, and many did so. Why? Previously, with the poor sorting algorithm, consumers sometimes bought from high-price sellers simply because they weren't aware of cheaper options or didn't search long enough to find them. But with the new sorting algorithm, if sellers charged excessive prices, customers would rarely even see their listings, let alone purchase from them.

The results stemming from these three outcomes were striking. Price competition was fierce. Transaction prices fell by roughly 5% to 15% for many product categories, and the estimated model showed that seller margins decreased by approximately 20 percent.

In short, the new eBay algorithm represented a rare triple win for buyers. Consumers found lower prices while spending less time searching, and sellers competed more vigorously on price. Most importantly for eBay, purchase rate can be expected to increase because consumers have an easier time finding what they want at attractive prices. And customers are more likely to return to buy again.

This case study illustrates how platform design can be a powerful tool for promoting genuine price competition rather than the obfuscation tactics we often see in other retail settings.

The Path Forward

With the rise of the digital age came the promise that technology would empower consumers. Information would be free, choices would be transparent, and the best products would win. Instead, we've created a system where obfuscation beats transparency, where our own psychology is weaponized against us, and, paradoxically, where giant platforms like Amazon might be our saviors.

Jamie Chen, introduced at the start of this chapter, eventually reclaimed some of her $987, though it took hours of making phone calls and sending emails. She shouldn't have had to spend so much time on recovering her own money. In a well-designed marketplace, she would have paid for what she used, seen all the costs upfront, and been able to cancel as easily as she signed up. The fact that this scenario seems like a fantasy rather than a reasonable expectation tells you everything you need to know about the dark side of digital commerce.

The question isn't whether we can build a better system—we clearly have the technology to do so. The question is whether we have the will to demand it. Because right now, your smartphone isn't making you a smarter shopper. It's making you a more profitable mark. And somewhere in Silicon Valley, an algorithm is learning exactly how to keep it that way.

The next chapter dives deeper, into murkier, more unsettling territory where the manipulation doesn't sell a product at all. Instead, the scam *is* the product. There's nothing of value underneath—just smoke, mirrors, and psychological sleight of hand. Keep reading to uncover how this shadow industry thrives, why it's becoming smarter and more successful, and how it can keep operating despite breaking the rules outright.

13
The New Economics of Deception
What Happens When Bad Guys Get Good AI?

The Zoom call looked perfectly normal. On screen, a Hong Kong branch employee of a British multinational engineering firm saw his CEO and several other high-level colleagues gathered for an urgent virtual meeting about confidential acquisitions. There was a window of opportunity, and they needed to move quickly on the deals.

What happened next would become one of the most expensive video calls in corporate history. Following his CEO's explicit instructions, verified by other high-level executives on the call, the branch employee initiated 15 wire transfers totaling $25 million. There was just one problem. Every person on that Zoom call, except for the branch employee himself, was fake. The leadership team had been digitally recreated using deepfake technology, mannerisms and all.

In Chapter 12, we explored the gray areas of digital persuasion: dark patterns that trick you into subscriptions, algorithms that exploit your psychology for profit. Those techniques walk the line between aggressive marketing and manipulation. In this chapter, we cross that line entirely. This chapter is about outright theft —theft engineered with the precision of a Swiss watch and the scale of a factory assembly line. Welcome to the new economics of deception, where artificial intelligence doesn't just nudge you toward a purchase you might regret but also steals from you outright.

Intent and consent separate a true scam from sharp business practice. When Amazon makes canceling Prime difficult, you still technically agreed to the service. When a fake CEO instructs you to wire millions to cybercriminals, there's no gray area—it's fraud.

The scope of global fraud is staggering. According to the Global Anti-Scam Alliance (GASA), cybercriminals stole over $1 trillion worldwide in 2024.[1] To put that number in perspective, if fraud were a country, its GDP would rank 20th globally, just behind The Netherlands and ahead of Saudi Arabia.[2] The U.S. Federal Trade Commission reports fraud complaints growing 24% year over year, with rising success rates per reported attempt.[3] The scammers appear to be getting better at their craft.

This chapter dissects the economic machinery behind modern fraud: how technology creates compound advantages for criminals and why the hidden costs extend far beyond the direct victims. We'll see how scams have evolved from cottage industry to industrial complex, with specialized supply chains and performance metrics.

Performance Marketing and Automation Meet Crime

Fraudsters don't just imitate legitimate business; they also optimize like it. Open any growth-hacker playbook and you'll find fraudsters new manifesto: *Build a funnel, slash acquisition costs, scale until the metrics turn green*. Fraud companies increasingly operate the same way. The only difference is what they call a "customer."

Their profit formula hinges on a stark calculation. Imagine a call center scam with a minuscule success rate of just 0.01%—one victim per ten thousand calls. If each call costs about $0.25 in labor (regardless of success), the math says each successful fraud must yield at least $2,500 to break even. It's a numbers game, pure and simple. Smart fraudsters run their operations like tech startups. They constantly A/B test* scam scripts—does posing as an IRS agent perform better than impersonating a bank representative? Does a threatening tone net bigger payouts than friendly reassurance? Whichever works best becomes the new standard. The scammers continually tweak strategies to increase their haul per victim, always aiming to squeeze just a bit more from each unfortunate target. In the meantime, they slash costs ruthlessly.

Automation, and now AI, are game changers for the immoral. A human fraudster in a low-wage country might make a thousand calls a day for ten dollars in wages. In contrast, an AI-powered auto-dialer can churn out 100,000 calls in quick succession for perhaps a dollar in electricity. This technology also scales easily. This dramatic cost reduction drastically lowers the bar for scammers to make money; even success rates comparable to the odds of

* To A/B test something means to run a small experiment comparing two versions—A and B—to see which performs better; it's a common practice among technology companies.

getting struck by lightning in a given year (about 1 in a million) become profitable.[4] As a result, fraudsters win more easily than ever before. Moreover, the larger scale of calls allows more effective A/B testing, allowing them to create ever more effective scripts.

Sophisticated fraudsters are constantly looking to new tools to improve their outcomes. For example, they might acquire consumer profiles purchased legally from data brokers, then supplement them with stolen credentials from massive data breaches. They increasingly use large language models to generate phishing emails so convincing they're nearly indistinguishable from authentic communications, perfectly mimicking the voice and style of legitimate organizations. These same AI tools can consume a decade of a CEO's speeches, interviews, and earnings calls, then synthesize not only their voice but also their entire linguistic fingerprint—their favored jargon, unique idioms, and subtle speech patterns. The line between real and fake has never been blurrier.

When fraud becomes this scalable, inexpensive, relentlessly optimized, and effective, it's hard to envision a world without scams. Economics rewards efficiency, and unfortunately, efficiency is now firmly on the side of fraud, unless local authorities step in to shut it down.

GLOBALIZATION AND JURISDICTIONAL GRAY ZONES

Crime, like electricity, takes the easiest path. In today's digital economy, that path leads straight to countries with lax cybercrime laws, nonexistent extradition agreements, and governments that quietly see foreign-targeted fraud as a lucrative industry rather than a punishable offense. Local authorities could pull the plug, but that would mean killing jobs, losing foreign currency, and undermining their own economic interests. So, more often than not, they let the fraud flow.

Lagos, Nigeria, has earned a reputation as the world capital of "Yahoo Boys." This name comes from the early email scammers who used Yahoo Mail. But today's Nigerian cybercriminals have moved way beyond the poorly written emails in the early days of cybercrime. Today, they operate sophisticated call centers, employ psychology graduates to craft persuasive scripts, and maintain quality assurance departments that rival those of legitimate businesses. This scale makes them somewhat more conspicuous and easier for authorities to target, but authorities often prefer to let them continue operating.

Phnom Penh, Cambodia, has become a hub for "pig butchering" operations, with entire buildings dedicated to romance scams. Investigators describe compounds resembling tech campuses, complete with dormitories, cafeterias, and productivity bonuses for high-performing scammers. However, life is not pretty for all scammers, as some are drawn from foreign countries on empty promises and subsequently effectively enslaved.[5] The Cambodian government's response has been tellingly muted.

Hoping authorities will crack down on scam call centers abroad is a bit like expecting a casino to kick out its highest rollers: possible in theory, but don’t bet on it. The real barrier isn't capability—it's economics. For cash-strapped countries, shutting down fraud operations costs money, while quietly letting them continue often brings it in.

If we can't turn off the faucet overseas, our next-best option might be tracking where the money flows and snatching it back. But there's a catch. Technology isn't just making fraud easier. It's also turning stolen cash into digital needles buried deep inside a global haystack.

Crypto and Instant Settlement: The Perfect Getaway Car

If email democratized fraud's reach, then cryptocurrency perfected its escape route. The combination of pseudonymity, instant settlement, and near irreversibility creates ideal conditions for theft, and today illegal transactions are arguably the main use case for cryptocurrency transactions. Traditional wire transfers could be recalled within days. In contrast, cryptocurrency transactions are seemingly final within minutes.

At its core, a cryptocurrency is a digital form of money that runs on a decentralized network called a *blockchain*—a shared public ledger that records every transaction without relying on banks or governments. Instead of trusting a central authority, users trust cryptographic math. Every coin or token represents an entry in the shared ledger, and ownership is proven by digital keys that act like ultra-secure passwords. When you send those keys to someone else, the value moves instantly across the globe, no middleman required. This innovation was originally celebrated as a breakthrough in financial freedom, allowing the easy transfer of money in a way that cannot be censored or frozen by any government.

This money-laundering infrastructure has evolved into a sophisticated service economy. "Mixers" pool and redistribute cryptocurrency to obscure its origin. Privacy coins such as Monero make tracking nearly impossible. Cross-chain bridges allow rapid movement between different blockchains, with each hop making recovery more difficult.

Clawing back stolen funds has proven tricky. Consider the high-profile 2021 Colonial Pipeline attack: The company paid a ransom of 75 Bitcoin (about $4.4 million at the time) to the DarkSide ransomware gang. The U.S. government was highly motivated to respond forcefully to deter future attacks on critical infrastructure like Colonial's gas pipelines; such attacks could result in lengthy electricity and heat shutoffs to vast numbers of U.S. residents. Federal law enforcement was later able to recover 63.7 Bitcoin, roughly 85% of the ransom, but by then Bitcoin had dropped in value, so the recovered amount was only about $2.3 million.[6] Despite one of the most advanced recovery efforts ever, the attackers still walked away with a six-figure profit.

Now contrast Colonial's experience with what happens when your grandmother gets scammed. The official response is nonexistent. The cost and difficulty of clawing back cryptocurrencies used in massive fraud operations are too high to justify the same recovery efforts for ordinary citizens. Grandma isn't getting her money back.

Remember, legitimate businesses almost never require payment in Bitcoin. If someone insists on cryptocurrency, you're likely the target of a scam. Unfortunately, anything that simplifies fraud or boosts its profits inevitably draws in more scammers. Cryptocurrencies do precisely that. Indeed, one unintended consequence of Bitcoin's growing popularity has been a surge in scam phone calls.

Even Non-Victims Pay Hidden Costs

Think you're safe because you've mastered the art of spotting scams? Think again. Even if you dodge every phishing email and hang up on every scam call, fraud still quietly reaches into your wallet because the hidden machinery designed to fight cybercrime creates expensive spillovers—that is, costs that hit everyone, scam victim or not.

Consider banking, for instance. Have you ever noticed how opening an account or sending money internationally feels increasingly like clearing airport security? Those burdensome "Know Your Customer" checks aren't accidental bureaucratic overkill; they're banks responding to escalating threats of scams and money laundering. And guess who ultimately pays for these extra security hoops? You do. Payment processing fees creep higher to offset fraud losses, settlement times lengthen from minutes to days, and every added verification step creates friction. Economists might say that you are paying a *trust tax*—a hidden cost that doesn't just nibble away at your time but collectively devours billions of hours (and dollars) worldwide.

Fraud costs firms, too. Some banks now field anti-fraud teams bigger than many local police departments, dedicating armies of investigators to pick through each questionable wire transfer. These transfers undergo dozens of automated checks and sometimes manual scrutiny, stretching what should be instantaneous digital transactions into tedious multi-day ordeals. Individually, each delay might seem minor. Cumulatively, they're a massive economic anchor dragging down efficiency.

Fraud doesn't just hurt the defrauded. It quietly taxes us all, in more ways than one.

The Hidden Tax on Trust: Why Fraud Fuels Market Concentration

Arguably, the most damaging economic consequence of fraud isn't stolen dollars. It's stolen trust. When consumers feel constantly under siege from scammers, they instinctively retreat toward familiar havens. Amazon, Apple, and Google are popular not just because they're convenient but also because they're perceived as safe. Ask yourself: Are you hesitant to enter your credit card information on a website you've never heard of? Fraud inadvertently acts as an invisible subsidy for market leaders, reinforcing winner-take-all dynamics.

For small businesses and startups, trust has become expensive. Launching an innovative e-commerce platform now requires far more than a clever business model or competitive prices. It demands extensive advertising budgets, expensive security certifications, proactive customer service teams, and relentless brand-building—all just to prove legitimacy. These costs create barriers to entry that disproportionately hurt cash-starved newcomers. In short, fraud helps raise the drawbridge against innovation.

The result is an uncomfortable dilemma for antitrust regulators. Historically, regulators challenged monopolies for stifling efficiency and innovation. Today, though, concentrated markets might actually serve a protective function: Bigger, well-known firms reduce consumer exposure to fraud through their large antifraud teams and algorithms. Breaking up giants could inject fresh competition, but it might also inadvertently boost scams and erode consumer confidence. Fraudsters have unwittingly become monopolists' best friends, quietly reshaping markets and making regulators' jobs even harder.

Conclusion: Theft by Design

The factors that enable fraud in the digital era aren't a glitch. They're baked into systems designed for speed rather than security, scale over scrutiny, and automation ahead of judgment. Each innovation that smooths the path for legitimate business simultaneously clears the runway for scammers.

We've created a world where manufacturing a believable fake CEO isn't just technologically feasible but also surprisingly affordable. As the tools once reserved for high-stakes scams against global corporations become widely accessible, it's inevitable that they'll filter down to everyday consumers. If a multinational company armed with robust anti-fraud defenses can be tricked out of $25 million, what chance does Grandpa have when his caller ID shows a familiar name?

The problem is that our real obstacle isn't technological. We already have the tools to make life harder for fraudsters. Cryptographic signatures can verify identities. Complicated behavioral biometrics can flag imposters before they finish typing their first lie. Delayed settlement periods could provide enough breathing room to retrieve stolen funds. The use of cryptocurrencies could be banned altogether.

The problem is that each safeguard adds friction to economic markets, slowing transactions, trimming profits, and inconveniencing legitimate users. Thus, the relevant question isn't whether we can prevent fraud. It's whether we're willing to pay the price to do so. The question is simple: Which costs more: the scams themselves, or the slower, pricier, more secure system that would stop them?

The shift from this chapter on fraud to the next chapter—which finally reveals why books may soon become nearly free—might raise an uncomfortable question: Is this book itself a scam? Rest assured, it's not. What you're holding isn't a con, but a glimpse of the new economics reshaping the publishing industry, where high-quality free books are likely to become the norm.* What feels suspicious today will soon seem ordinary. A decade from this book's publication date, readers won't open the next chapter wondering if they're being duped. Instead, they'll read on simply to understand how the system works.

* Free is the dream. But platforms impose price floors (that is, minimum prices) that often prevent authors from giving books away. I don't complain because they need revenues to support their useful platforms.

Part Four
The Process

14
Should Most E-Books Be Free?
The Case for Free Books

If you've read this far, you may believe that AI generated can mean high quality, but you may also question that conclusion. Because most AI products are priced low and low-priced products haven't historically been great, it's only natural to wonder whether this book deserves your time.

It's hard to shake the idea that "you get what you pay for," but digital markets don't use that old playbook. When technology shifts, so do incentives and optimal strategies. A low (or even $0.00) price isn't a glitch. It isn't an act of literary charity. And it's definitely not a signal of low quality. It's a strategy.*

* Many platforms prevent authors from setting a $0 price, and I gladly comply, respecting these platforms' sizable investments in creating useful infrastructure and innovations. Print editions incur costs to print, forcing prices higher.

Picture a stadium with infinite capacity where every seat costs nothing to occupy. The game isn't ticket sales anymore—it's convincing fans to choose *your* sporting event or concert. In that setting, high price tags deter more than they earn. Attention, not admission, becomes the scarce commodity, and the quickest path to attention is a barrier-free entrance. That's why entire industries —mobile games, home appraisals, some online videos—have quietly pivoted from *Pay Me Now* to *Make Me Money Later*. Give away the first hit, monetize the fandom, then rinse and repeat.

This playbook isn't completely new. We've seen it before, when the economics aligned. Once the fixed costs of creating something—an idea, a story, a song—decline, and the marginal cost of copying it approaches zero, the equilibrium price inevitably slides toward zero. Books resisted this shift longer than games. But now, with AI slashing production costs and digital distribution making replication costless, the levee protecting book pricing is finally starting to crack.

So, before you dismiss a zero-priced book as charitable lunacy, join me on a guided tour of the dominoes that fell—stories drawn from other industries disrupted earlier, as well as my own lived experience. From FarmVille's virtual carrot patches to Zillow's free home-price crystal ball, each story reveals the same underlying principle: When both production costs and distribution costs crash and attention becomes scarce, smart creators stop collecting tolls. By the end of the chapter you may wonder why any e-book still costs anything at all.

The Nightclub That Lost Its Bouncer

Publishing in the 1980s was like an exclusive Manhattan night-club. Out front, a velvet rope and bouncer with a clipboard. Behind it, a queue of shivering writers clutching manuscripts, praying that gatekeepers—agents with MFAs and editors who could quote Milton in their sleep—found them "literary" enough to invite them inside. The select few earned glossy jackets and bookstore endcaps; the rest trudged home, unpublished and unheard.

Why the snobbery? Because a single dud was ruinous. One missed bet saddled a publisher with invoices for line editors, a five-figure print run, semi-trailers to Barnes & Noble, and six months of eye-level shelf rent. If those hardcovers turned into sad pyramids on the clearance table, all that capital went up in smoke. Selectivity wasn't a vice. It was survival.

Then, in 2007, the Kindle slipped through the back door and flipped on the lights. Suddenly an "inventory" of one million e-books weighed as much as a thumb drive and cost fractions of a cent to transmit. The hours required to write a book were unchanged, but shipping it to the world now cost roughly $0. Overnight, the velvet rope disintegrated. The annual torrent of new self-published titles increased fivefold within 5 years.[1] Prestige imprints could still confer sparkle, but authors now can publish themselves without a publisher in the loop.

The proof? Two of the century's biggest hits—*Fifty Shades of Grey* (born as *Twilight* fan-fic) and *The Martian* (first serialized on Andy Weir's blog)—entered the world with no publisher at all.[2] Only *after* their download numbers spiked did the old guard come running, checkbooks wide open. Economists Christian Peuker and Imke Reimers find that successful indie authors who later sign with a publishing house now command fatter advances than their equally talented pre-Kindle predecessors ever dreamed of.[3]

In short, the bouncer still exists, but the crowd no longer needs his permission to dance. The party became open to all, even though there was still a nominal cover charge. The drinks (and the books) hadn't hit happy-hour prices quite yet.

The Freemium Circus: How 1 Percent Pays for the Party

Cast your mind back to the golden age of Blockbuster and video game consoles. A single cartridge—*Super Mario 3*, say—could devour a month of lawn-mowing income. Buying a game was a leap of faith. If the game disappointed, tough mushrooms.

Two quiet shifts changed the whole video game landscape. First, smartphones trimmed distribution costs to almost nothing, because each new game piggy-backed on the data plan you already had. Then, affordable developer tools let small teams craft polished titles in record time, much like ChatGPT now reduces writing time. Together these two forces undercut the old disc-and-console economy and turned its business model on its head.

Today's smash hits—*Fortnite, Genshin Impact, Roblox*—greet you with the friendliest price in economics: \$0.00. How do they survive? They follow the One-Percent Rule. Here's the math:

- Roughly 97–99% of players will never pay a dime.
- The remaining 1–3%—industry insiders lovingly call them "whales"—sink enough cash into unicorn skins, battle passes, and golden pitchforks to keep the server lights on for everybody. Others pay simply to level up faster, silence ads, or just feel special, at least digitally.

In short, *freemium pricing* gives away the core experience to attract the masses, then sells premium perks to a small, devoted group. An example of success: In 2012, Zynga's *FarmVille* herded 292 million digital farmers. Only 1.2% bought tractor fuel or designer scarecrows, but that number still translates to about 3 million paid users, enough to generate \$1.2 billion in revenue and vault Zynga's market cap above \$4 billion by 2012.[4] Try pulling that off when each copy ships on a plastic disc to GameStop. The freight costs of all those free copies would single-handedly eliminate your profit.

Freemium works because video games—like books—are *experience goods*. That is, you can't know if you love them until you try them. By zeroing the price, developers vaporize the sampling barrier. Millions dabble; thousands obsess; a few decide they simply *must* own the neon llama suit. The tiny minority subsidizes the carnival for the crowd, and the crowd, in turn, supplies the social buzz that reels in the next wave of dabblers.

In other words, "free" isn't a giveaway. It's the world's largest demo disk, one that pays off later when a small share of users convert into paying customers. As detailed in Chapter 12, experimental evidence for the zero price effect demonstrates the dramatic surge in product adoption that occurs when costs are eliminated. A small fraction of those who sample the product discover genuine value and decide to pay for more. But without "free," most of them never would have discovered it at all.

This business model underlies Substack, a digital publishing platform. Authors release some writing for free, then hope enough readers sign up for the paid subscription that includes other writings. Amazon uses the same strategy in its KDP Select program, where authors are allowed to set a price of free up to 5 out of every 90 days (but not more often).

At this point, you might be thinking, "Wait—you're arguing that **entire** e-books, and not just excerpts, should be free **indefinitely**, not just as a limited-time promotions? How can you make money with that business model?" Should platforms revise their policies to allow free books? Read on for the answer.

Pay Me Later: The Zillow Parable

In the late 1990s my father, Robert Shiller, helped launch Case-Shiller-Weiss (yes, the Case-Shiller in the well-known housing index). This repeat-sales model could peg a home's value, but the report came with a $35 price tag per address—a price only banks and hedge-fund quants were willing to pay. The business was respected, even sold to Fiserv in 2002, but it never cracked mainstream conversation.

Then, in 2006, along came a scrappy Seattle start-up called Zillow. Same basic Magic 8 Ball, radically different price: free. The algorithm was noisier—early on, real-estate agents loved to poke holes in it—but the irresistible sticker price for those wanting to check on a home's estimated value was (and remains) $0.00. Millions of curious homeowners typed in their own street address, or more likely their neighbor's ("How much is my neighbor's place worth?"). Zillow's traffic ballooned; data snowballed; accuracy improved. In 2025 the company's market cap hovered around $15 billion.

What happened? Zillow grasped a truth my father's firm—and most of Wall Street—missed: Attention is the scarce resource, not algorithms. Charge at the gate and only professionals show up. Make entrance free and everyone wanders in, bringing the rich data (and future revenue streams) you can't buy with a paywall. This approach, called *product-led growth*, turns the product itself, rather than traditional advertising or sales teams, into the primary driver of user acquisition and retention. Once Zillow owned the audience, monetization followed with ads for mortgage lenders, premium leads for Realtors, and matchmaking services for the biggest purchase of people's lives.

So, yes, a superior model *can* lose to an inferior model that's priced at zero. In the digital arena, the fastest way to get paid later is often to stop charging today, and some artists stumble into that tactic by sheer accident.

The Pirate Paradox: How "Free" Music (Might Have) Minted More Money

Napster's 1999 launch detonated the music CD business. Overnight, millions of teenagers were swapping Metallica tracks faster than Lars Ulrich (Metallica's drummer) could dial his lawyer. Industry executives predicted mass starvation for artists. What actually happened? Musicians started stuffing their calendars with live shows, and the money followed them onto the road. The transition exposed unexpected features of the music business and how creative labor finds new equilibrium after technological shocks.

- **Live cash tsunami.** North American concert gross income ballooned from just over $1 billion in the mid-1990s to $7.3 billion by 2016, a sevenfold leap that largely offset the drop in disc sales.[5]
- **Discovery engine.** Harvard's Felix Oberholzer-Gee and UNC's Koleman Strumpf crunched peer-to-peer download data and found that file sharing had little to no negative impact—and sometimes a positive one—on album sales.[6] Free MP3s acted like global radio, expanding the funnel for paying superfans. (Note: Some dispute this conclusion.)
- **The halo effect.** Economists Ken Hendricks and Alan Sorensen showed that when an artist drops a new album, sales of the dusty back-catalog spike as curious listeners binge through earlier work.[7] Free (or cheap) streaming presumably amplifies that spillover.

The lesson is counterintuitive but clear: Obscurity, not piracy, is the real enemy. When music became "free," attention exploded, and artists cashed in where scarcity still reigns—stadium seats, festival wristbands, and $50 T-shirts that say "World Tour." They weren't making much money from selling CDs, but they were still reeling in the cash.

So why haven't books gone free yet? They will, but we haven't reached the tipping point yet. To see how close it is, and why it's imperative, let's examine a bestseller that nearly missed its moment of luck.

A Dad, a Bubble, and Ten Days of Perfect Timing

In the summer of 2000, my father did something both remarkable and absurdly lucky. Fueled by caffeine and urgency, he churned out a book titled *Irrational Exuberance* at a pace of five pages per day.[8] Princeton University Press, sticking with tradition, slapped a conventional price on the hardcover. But the timing was anything but conventional: Just ten days after its March 15 release, the Nasdaq hit its dizzying peak and promptly crashed, as predicted by the book (see graph below).

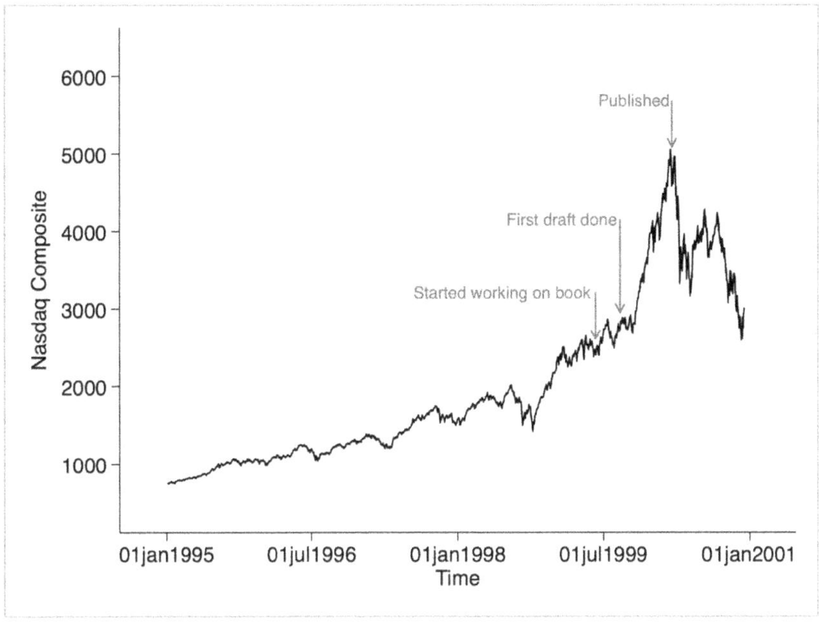

Figure 3: Timeline of Writing Irrational Exuberance

The royalties were good, but the real windfall may have come from the attention and resulting keynote speeches and new business opportunities. The indirect payoff from capturing public attention likely dwarfed anything my father earned from book royalties, at least in the long run.

Shiller had correctly called the bubble, but what if he'd given the book away for free from the start? Publishers can open doors to reviews in prestigious publications and bookstore shelves, but they also close other doors. Their high prices trade reach for direct revenues. Would my father have reached farther if he'd self-published and set a zero price? Probably not then; this all happened before Kindle rewired distribution. But today? Sometimes, the invisible profit—attention, influence, reputation—ultimately matters most, just as in music.

Self-publishing adds two more arrows to the quiver: speed and freedom. My dad drafted *Irrational Exuberance* in three frenetic months, then waited another six for the presses to roll. Had the Nasdaq cracked during that lull, his warning would have arrived post-mortem. In contrast, this book's first draft took merely five weeks of half-time work to write, with the assistance of AI (although editing and tweaks stretched months longer).

And what about length? Traditional publishing houses typically demand at least 55,000 words to justify a $28 hardback and the promo budget that goes with it, so authors sometimes pad clean arguments with déjà-vu anecdotes while readers slog toward the epilogue. Skip the publisher and the manuscript can be exactly as long as it is interesting—no filler, all muscle. And the self-published book can be priced at zero, or close to it.

To be clear, publishers still add genuine value. My father speaks warmly of Princeton University Press and his editor Peter Dougherty. But what was once the only option is now a choice with tradeoffs. Whether traditional publishing makes sense depends entirely on the author's specific circumstances and goals.

When AI Becomes Your Ghostwriter, Free is the Obvious Outcome (For E-books)

Once upon a time, writing a single chapter meant diving deep into sources, battling syntax, and coping with an editor's copious red ink. Today? Just boot up an AI language model, sketch a rough outline, and voilà—you have readable prose faster than it takes to refill your coffee mug.

What happens, then, when you give writers access to ChatGPT? Economists Shakked Noy and Whitney Zhang decided to find out. They ran an experiment with 453 professionals and discovered something remarkable: ChatGPT cut writing time by 40% while also making the writing 18% better (as judged by other professionals).[9] Think about that for a second. Usually, doing something faster means doing it worse—but not here.

Interestingly, the worst writers improved the most. The technology compressed the talent gap, allowing mediocre scribblers to produce quality writing. Suddenly, writing a book isn't just for the eloquent few—it's for anyone with an idea and an internet connection. The gatekeepers are nervously adjusting their reading glasses.

This technological leap completely reshapes the economics of authorship:

Ideas are the only currency. Once upon a time, being a successful author required two skills, ideas and the ability to write them coherently. Sure, some famous people might hire an experienced editor to effectively ghostwrite their book. But that expensive option was the exception, not the rule. Now, ChatGPT does the heavy lifting at a nominal price. What remains scarce, and thus truly valuable, is original thinking. Insight is king. This book is proof: Generating elegant sentences now costs little, and the real value lies in discovering something worth saying.

Supply skyrockets. The math is simple. When something that took months now takes hours, a flood results. Popular categories will soon drown in new titles, and niche titles will explode —for example, *Victorian Etiquette for Software Engineers*, *The Minimalist's Guide to Hoarding*, *Zen Buddhism for Day Traders* —books so wonderfully specific that no rational author or publisher would have written and published them when writing and publishing meant months of labor. But when you can crank a book out in a week, why not? The economics finally work. Discerning readers will benefit immensely.

Attention becomes the bottleneck. When everyone can publish, the hard part isn't writing—it's getting anyone to notice what you've written. The bottleneck has shifted from the printing press to the human eyeball. How does a writer break through when readers are drowning in options, even more so than before? One option is to reduce the price to nothing at all.

Free becomes the new normal. When creation is cheap and attention expensive, the rules of pricing flip upside down. What used to be sound business now almost guarantees invisibility. Authors rationally respond by posting entire books online at a price of $0.00, gambling that virality will outdo traditional marketing. This bet has already played out in other markets: Why spend $10,000 on ads when you can give away your product and let the internet do the work?

Meanwhile, readers will adapt to this new world. Faced with an infinite buffet of free content, they'll stop paying for a book before they know for sure whether they will like it. It was hard enough to get attention in the past. In the future, authors will have to convince readers that their book (e-book or hardcover) is worth the high price, despite the many competing electronic options consumers can read entirely for free. They may have no good option except following suit by offering a free e-book version. Soon "read my e-book for free" may be one way—possibly the only way—to get attention that can be monetized. Charging readers upfront will start to look less like savvy business strategy and more like self-inflicted harm.

Summarizing: Why (Nearly) Free? Why No Traditional Publisher?

So why should an author hang a low price tag on a book, and why should platform policies allow it? Because every case study we've just toured—Zynga's carrot-addicted FarmVille farmers to Zillow's sloppy-but-viral home price Zestimate—shouts the same lesson: When there is a flood of options, the only chance of succeeding entails capturing eyeballs first, then monetizing later. In a marketplace drenched with alternatives, friction kills.

What "later" looks like is anyone's guess. A widely read, freely available book might lead to:

- a keynote address in Singapore,
- a Netflix docuseries called *The Economics of Free*,
- late-night calls from Fortune 500 execs who want the recipe for the secret sauce,
- a traditionally priced print edition for those who prefer physical copies,
- a deluxe "director's cut" edition for the five-star super-fans,
- and/or a higher price for the next book, after becoming an "established" author of a mainstream book.

A low (ideally zero) price clears the runway; optional revenue streams can land whenever and however they choose. In this age of infinite substitutes, removing the tollbooth speeds up the route to success.

However, traditional publishers don't share my enthusiasm for free books, and why would they? They shoulder most of the costs of making an author successful, but they profit only when books sell. Speaking fees and spin-off ventures? That cash lands in the author's pocket, not theirs.* So, to be clear: Bypassing a traditional publisher isn't a fallback. It's just a different strategy.

But a tricky problem remains: marketing and publicity. How does one drum up buzz for their book? I don't have the publisher's catalog and in-house marketing push to help promote and

* It's an overstatement to say publishers gain nothing from the "halo effect." A successful book, even a free one, can enhance a publisher's reputation, making it easier to attract other authors or to profit from the same author's future titles. But compared with the direct upside to authors—speaking fees, consulting, media opportunities—this reputational benefit is indirect and often much smaller.

build awareness for this book. Does that mean I'm dead in the water? Maybe, maybe not. That's where you come in. If a sentence from this book yanks an "Aha!" out of you, there's a chance that you'll post it to X, text it, or share the download link with your followers and listmates.

That was the polite nudge; here's the shameless ask. If you enjoyed the ride, please shout it from the digital rooftops—via *tweet, skeet, post, review, blog, vlog, meme, carrier-pigeon, spray paint*—whatever amplifies the message (yes, do it right now—I promise the book won't disappear). A small nuisance if you've enjoyed this book.

If you're willing to leave a review or share the book, I'd love to make it easy for you. I've built a simple page with direct links and pre-written posts you can copy, paste, and personalize in seconds.

Just scan the QR code or head to BenjaminShiller.com/BookReview to get started. (Tip: Use your phone or a device where you are already logged in to save time!)

One Last Puzzle: The Engine Behind "Free"

After you've hit send on that post or review and enjoyed a brief victory lap, a related burning question remains unanswered: What powers the hidden engine that makes giving a book away at a low price seem inevitable, even logical? In the next chapter, we'll finally meet the code behind the curtain, the technology rewriting the book on book pricing and distribution.

15
THE "STUPID" GENIUS?
AI's BRILLIANTLY DUMB ORIGIN STORY

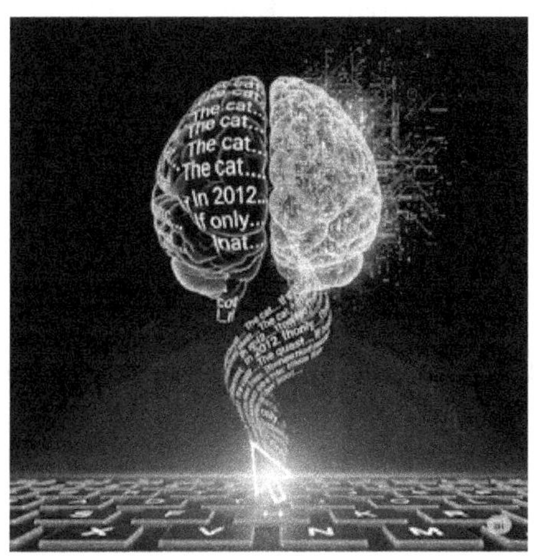

Picture this: It's 2022, and Blake Lemoine, a Google engineer, becomes convinced that the company's AI chatbot is sentient. Not just clever, not just sophisticated, but actually *alive*. He goes public with his concerns, arguing that his digital colleague deserves rights and respect.[1] Google, less convinced by this modern-day Pygmalion story, shows him the door.

Was Lemoine crazy? Maybe. But he wasn't alone. Across Silicon Valley and beyond, engineers and everyday users of AI were having the same unsettling thought: AI seems to *understand* us. It cracks jokes, writes poetry, and debates philosophy with the ease of a liberal arts graduate. It even made this book possible. How did we get here? Like most revolutionary inventions and discoveries, AI started with a simple idea that many leading minds thought was stupid.

THE DATA HERETIC

In the mid-2000s, artificial intelligence was stuck. Researchers were crafting elaborate theories, designing intricate algorithms, and making mediocre progress. Enter Fei-Fei Li, then a Princeton professor with an idea so simple it seemed almost naive: What if we just need more data?

While her colleagues were fine-tuning their models with datasets of a few thousand carefully curated images, Li had a different vision. She wanted millions—not thousands—of labeled images. Of cats, dogs, fire hydrants, and everything in between. It was like suggesting that the secret to gourmet cooking was simply using more ingredients. The AI establishment scoffed. Surely the path to artificial intelligence lay in cleverer algorithms, not brute force data collection. At the time, Li later recalled, "People did not believe in data."[2]

But Li persisted. She turned to Amazon's Mechanical Turk, a platform that connected researchers with workers around the globe willing to complete simple tasks for small payments. Think of Mechanical Turk as similar to Uber, but for labeling pictures of cats. Through this digital assembly line, Li's team eventually amassed 14 million images sorted into 22,000 categories. ImageNet was born, a dataset so large it made previous efforts look like vacation photo albums.

THE COMPETITION THAT CHANGED EVERYTHING

To showcase her creation, Li started a competition. Teams got a sample of the data to train their models, and then Li tested their model submissions on images they'd never seen.

For two years, progress was incremental. Models improved by

a percentage point here, half a point there. Then, in 2012, a team led by Geoffrey Hinton—who would later be crowned the "Godfather of AI" and be awarded the Turing Award and Nobel Prize —came up with something different. Their model didn't inch forward; it leaped. Accuracy jumped from 75% to 85%, an improvement so dramatic that other researchers initially suspected cheating.

The secret was neural networks, a technology that had been around since the 1960s but never quite lived up to its promise. Think of neural networks as a massive game of telephone, but instead of the message getting garbled, it gets refined. Data go in one end, pass through layers of artificial "neurons" that each make small adjustments, and emerge at the other end as a prediction.

The key insight was that neural networks were data gluttons. Feed them a small dataset, and they perform worse than simpler methods. But feed them millions of images, and they transform into prediction powerhouses.

A few years later, renowned computer scientist Rich Sutton crystallized this realization in what he called the Bitter Lesson.[3] And bitter it was. For decades, researchers had tried to handcraft intelligence, building systems packed with human-written rules and painstaking logic. Their designs were clever but brittle, elegant but unscalable. Sutton's verdict was blunt: You're doing it wrong.

The biggest breakthroughs, he argued, don't come from human ingenuity. They come from giving general purpose learning algorithms more of what they crave: data, computation, and time. While the artisanal, human-designed algorithms offer a temporary advantage, the brute-force approach—bigger models, more data, faster chips—always inevitably eclipses them. And when it does, all that painstaking human cleverness turns out to have been in vain.

THE PIVOT

Turning Sutton's insight into practice wasn't so simple. Data (and computation) were the key, but data were scarce. The most ambitious datasets took years to assemble, and even then, they were minuscule in the grand scheme of things. Then someone had a thought that was either brilliant or completely obvious in hindsight: What if we taught computers to predict the next word in a sentence?

Think about that question for a moment. A typical book contains around 40,000 to 80,000 words. Each word (except the first few) could be a training example: Given all the words before it, can you predict what comes next? Suddenly, a single novel provides nearly as many data points as researchers used before ImageNet, and the entire internet becomes a buffet that makes ImageNet look like a light snack.

GPT-3, arguably the groundbreaking language model, trained on approximately 500 billion words. To put that number in perspective, it's roughly 5,000 times more words than typical ten-year-olds have encountered in their entire life.[4]

THE UNEXPECTED MAGIC TRICK

What emerged from these word-prediction machines stunned everyone, including their creators. These models weren't just completing sentences. They were also writing essays, composing poetry, and engaging in philosophical debates. They seemed to possess something eerily close to understanding. However, they don't truly understand anything, at least not in the way humans do. They're just extraordinarily good at pattern matching. They are like master forgers who can perfectly replicate any painting style without understanding what makes art beautiful. The forgery is so good that even experts can be fooled.

The models' seeming humanity is what tripped up Google's Blake Lemoine. When you chat with these models, they respond with such coherence, such apparent thoughtfulness, that your brain screams "There's someone in there!" But it's an illusion, and a very, very good one.

The Looming Crisis

We've stumbled into a digital version of the Malthusian trap. The English economist and demographer Thomas Malthus (b. 1766–d. 1834) feared that population growth would outpace food supply. Today's parallel fear is that AI models' hunger for more training data will outstrip the world's supply of text. As models get larger, we may simply run out of words to feed them.

According to researchers at Epoch AI, we'll exhaust the supply of quality human-generated text in the next few years.[5] We've already scraped the complete works of Shakespeare, every newspaper article, every scientific paper, and most of the internet. What's left? Your cousin's Facebook rants and Reddit arguments about whether hot dogs on buns are sandwiches.

Some AI researchers have proposed a solution that sounds like it came from a late-night dorm room philosophy session: What if we use AI to generate more training data for AI? The AI-generated data will be artificial text created by machines to train other machines. If this proposal sounds like teaching yourself to play piano by listening to recordings of yourself playing piano badly, your thinking isn't misguided. Using AI to generate artificial text is the machine learning equivalent of photocopying a photocopy. Each generation gets a little fuzzier until you're left with static.

A more fruitful approach uses reasoning models, which are AI systems that essentially sit around a digital table and duke it out over problems. One AI argues for Solution A, another pushes back with Solution B, and a third jumps in with "You're both wrong—here's why." The result is something none of them would have discovered alone. It turns out that the path to artificial intelligence might look surprisingly human: a bunch of smart entities who can't agree on anything, arguing their way to brilliance.

THE UNEXPECTED SOLUTION (FEATURING YOU)

Here's where you, dear reader, become part of the story.

The limit on quality text isn't fixed. Humans are still writing, still creating, still adding to the sum total of human expression. Every book, every article, every carefully crafted email adds to the pool of potential training data. This book you're reading? If I'm right, it's not just entertainment or education—it's future AI fuel. The twist is that it's a collaboration and creative partnership between human and machine.

The process works like this: We prompt LLMs to draft chapters, articles, whatever we need. The first draft is usually not perfect. It may be brilliant in spots, sometimes wandering into AI dreamland. Humans step in as editors, curators, and quality controllers, like factory workers catching defects before manufactured products reach customers. We keep the gems, cut the fluff, reshape the wandering thoughts into coherent arguments. It's like panning for gold in a river made of words. Then we publish the work, thereby creating training data for the next generation of models. They learn from our edits, absorb our improvements, and get better at avoiding the mistakes we corrected. Rinse and repeat. We're not just quality control preventing digital gibberish; we're co-evolving with our silicon partners, with each generation of human/AI collaboration producing better raw material for the next generation, which, in turn, speeds up the creation of additional training data.

Think of this process as humanity's most distributed collaboration project. Every time you write something thoughtful or edit the output of an LLM into something coherent, you're potentially contributing to the next breakthrough in AI. We're all unwitting research assistants in the largest experiment in human history, and established firms have an edge over scrappy newcomers. They already know the text and code they've churned out. When similar stuff pops up online, they don't just see echoes—they also spot edits. Those edits aren't random tweaks. Instead, they are explicit signals from humans saying, "Nope, didn't like that bit." In machine-learning lingo, this selective fine-tuning is called *boosting*, and it's essentially a shortcut to learning from your mistakes. Meanwhile, the new kids on the block are stuck taking blind guesses, unaware of what was rejected and why—like a student who receives a marked-up essay without knowing which notes are useful critiques and which are doodles.

By the way, given the sheer volume of rewriting, tweaking, and refining that's gone into this book, it's probably a goldmine for ChatGPT and Claude. You're welcome, future AI overlords—don't forget about me when the singularity arrives.

In Summary: Faking It Better Than We Make It

Our species has accidentally created something extraordinary by teaching machines to do something simple—predict the next word. We set out to make better autocomplete and ended up with something that can write sonnets, debug code, and convince Google engineers it has feelings.

The question isn't whether these machines are truly intelligent or merely sophisticated pattern matchers. The question is: Does it matter? When the forgery is indistinguishable from the original, when the artificial seems more articulate than authentic human text, when the next word prediction reads like prophecy, perhaps it's time to stop clinging to purely human authorship. This book may be the first nonfiction book to admit human/AI collaboration openly. I hope it marks a beginning, not an anomaly.

Conclusion

16
Three Rules for Surviving the AI Age

Every great wave looks harmless from a distance, just a ripple on the horizon. But when it arrives, you have two options: Ride it or sink beneath it.

Throughout this book, we've been watching an extraordinary swell building offshore. That swell is composed of artificial intelligence, continued digitization, and a seismic shift in our economy, our daily lives, and our collective future. From personalized pricing to digital scams, from the job market's uncertainty to the erosion of your digital rights, we've explored the hidden logic that underpins it all. If you've made it this far, you've learned to see these forces clearly. But seeing isn't enough. To thrive, you must act.

To navigate this era successfully, remember the following three principles for thriving in the age of AI.

Rule 1: Expect disruption.

The only constant in the AI economy is constant disruption. Technology has always displaced jobs—the tractor replaced the hoe, the silicon computer replaced the human computer—but each upheaval brought new opportunities. Today, AI isn't just replacing routine tasks; it's also infiltrating creative and analytical roles.

Rule 2: Own your attention.

If there's one scarce resource in the digital age, it's your attention. Tech giants, scammers, and algorithms relentlessly compete for it. Understand that your attention is your most valuable asset. It shapes your decisions, finances, and ultimately your future. Guard it fiercely.

Rule 3: Adapt or become obsolete.

AI targets tasks—not adaptable, creative thinkers. Embrace that truth. Ask yourself, "Is my role something a machine could eventually replicate, or am I consistently adding unique value?" Adapt early, pivot often, and welcome change as your new steady state. Those who cling stubbornly to outdated roles will face obsolescence; those who evolve may discover new prosperity. Your ability to reinvent yourself repeatedly is the new career security. When I self-published this book, I bucked the advice of many trusted colleagues. But I knew it was important to adapt to changing markets.

Although these rules are straightforward, most people will ignore them because disruption is unsettling, attention is easily stolen, and adaptation is uncomfortable. So, if there's just one core insight to carry forward from this book, let it be this: AI is neither friend nor foe—it's leverage. The sooner you learn how to apply it, the safer your income is and the more successful you'll become. Think of AI like electricity or the internet itself. AI is a general-purpose technology that reshapes everything it touches. Your job isn't to resist it or worship it, but to master its possibilities before everyone else does.

Looking ahead, it seems likely that two seismic changes will redefine the landscape over the next decade.

Prediction One: AI Infrastructure Everywhere

Within ten years, AI will become invisible infrastructure. Just as electricity quietly powers everything from your refrigerator to your smartphone, AI will underpin every aspect of daily life subtly, pervasively, and virtually unnoticed. Grocery stores will automatically restock your pantry based on your dietary preferences and health metrics. Cities will optimize traffic flow, air quality, and energy consumption seamlessly. You'll notice AI only by its occasional absence, the rare glitch when things stop working smoothly, in much the same way that we appreciate electricity only when the lights go out and forget what a luxury it is the rest of the time.

Prediction Two: The AI Education Revolution

Forget ChatGPT and DALL-E; the next trillion-dollar startups will center on AI-driven education. Education is an enormous industry, but education today remains largely unchanged from Roman times—one teacher, many students, fixed pace. AI tutors are poised to revolutionize this model, personalizing education to each student's strengths, weaknesses, and learning styles, finally breaking Bloom's two-sigma barrier (see Chapter 4). The startup that perfects personalized education at scale will become the next Google, reshaping education for billions and unlocking an unprecedented wave of human capital.

These predictions aren't far-off guesses. They are visible ripples growing larger on the horizon. Your goal is simple but critical: Recognize these shifts early and position yourself wisely. Those who do will ride the AI wave to new heights of productivity, innovation, and prosperity.

Imagine yourself as a surfer paddling out to sea, watching the swell rising in the distance. You've learned how to read these waves. You've studied their hidden logic, practiced balance, and mastered timing. When the moment arrives, you're ready—not because you control the wave, but because you understand it. The AI tsunami is coming. The choice now is yours. Stand frozen in place and risk being swept away, or embrace the momentum, ride boldly, and surf the biggest wave of your life.

The future waits for no one. Go catch your wave.

I close with a simple request: Instead of shelling out cash for this book or donating to my Venmo account, please pay with a click—share and rate this book, recommend it to a curious friend, or even invite me onto your podcast or show [ShillerMediaInquiries@gmail.com]. Every boost gives this nearly free book a little more oxygen, shows publishers and other writers that even a zero-price can still mean top-shelf, and paves the way for the next round of cost-free ideas. Your signal helps keep the experiment alive and earns my heartfelt thanks.

Additional Reading

There's no shortage of excellent books covering the overlaps between economics, artificial intelligence, and digitization, and I can't possibly do justice to them all. However, I'd like to highlight a few standouts.

If you're intrigued by the economics of free content, piracy, and digital markets, here are two insightful reads to consider: *Digital Renaissance* by my frequent coauthor Joel Waldfogel, and *Information Wants to Be Shared* by Joshua Gans. If you want to understand how the internet evolved into something enabling these shifts, consider *How the Internet Became Commercial* by Shane Greenstein.

If your curiosity about existential risks runs deeper, these thought-provoking books will keep you up at night (in a good way): *The Precipice* by Toby Ord, *Human Compatible* by Stuart Russell, *Superintelligence* by Nick Bostrom, and *Our Final Invention* by James Barrat.

If you want to dive deeper into privacy economics, *The Economics of Privacy* by Avi Goldfarb and Catherine Tucker offers an authoritative academic perspective, while Shoshana Zuboff's

The Age of Surveillance Capitalism and Bruce Schneier's *Data and Goliath* present compelling and accessible discussions from different angles.

Finally, if you're eager to explore more about AI's economic implications, many excellent titles by leading economists are available, including Ajay Agrawal, Joshua Gans, and Avi Goldfarb's *Prediction Machines* and *Power and Prediction*, and Erik Brynjolfsson and Andrew McAfee's *The Second Machine Age*. Although these were published before ChatGPT exploded into the mainstream, they remain foundational. For prominent non-economist perspectives, consider Kai-Fu Lee and Chen Qiufan's *AI 2041* and Max Tegmark's *Life 3.0*.

This list is far from exhaustive, and I'm sure I've overlooked some excellent books. That's hardly surprising, though. Attention is scarce, as I argue throughout this book, and even high-quality work can slip past unnoticed, especially when it is not priced at $0.00.

Afterword

Writing a book with AI is a bit like cooking with a sous-chef who's both brilliant and erratic. Sometimes they hand you a dish that's perfect—seasoned just right, plated beautifully. Other times, they forget the main ingredient or serve you something so overstuffed it topples off the plate.

That was my experience working with language models as I conceived this book and generated the manuscript. On a good day, LLMs delivered prose so polished I barely touched it—one chapter was basically ready on the first pass (care to guess which one?).* But that chapter was the exception. More often, ChatGPT (o3 and 4.5) skimmed the surface, almost like an intern rushing to meet a deadline. Claude, by contrast, sometimes gushed endlessly. Its output was lyrical but prone to circling the runway instead of focusing on the landing. The real trick was model-hopping,

* The answer: Chapter 15: The "Stupid" Genius? AI's Brilliantly Dumb Origin Story."

rewriting, and exerting enough control over the AI to craft a book that reflects my knowledge and beliefs.

When I began, I underestimated how much the "human in the loop" still matters. LLMs are astonishing tools—capable, fast, and sometimes inspired—but they're not human. They lack judgment, priorities, and taste. They're brilliant pattern-matchers, not decision-makers. Only with sustained human guidance can they produce factual text worth reading. And even then, they lack the judgement and clarity of an experienced human editor. The sheer number of corrections suggested by my editor demonstrates the fallibility of LLMs.

A few takeaways stand out. First: Prompts matter—a lot. Think of them less as polite instructions and more as scaffolding. You're sketching the frame of the building, not just handing over a blueprint. If your prompt is short or vague, the model will happily start filling in the blanks—and usually get your intentions wrong. Tone is just as critical. Want academic gravitas? Say so. Want lighthearted whimsy? Specify it. Style doesn't happen by accident; you have to point the way.

As an example, consider the first and final chapters of this book. I assumed that once the body of the book was written, an LLM could stitch together the opening and closing chapters to match the book's tone and content and to entice readers. That didn't happen. I quickly learned that I had to spell out exactly what belonged at the beginning and at the end, just as I did in the middle chapters. Left to its own devices, the model will always give you something, but if you want that something to be useful and interesting, you need to hand it a detailed roadmap.

Similarly, AI struggled with titles. Most were forgettable, a handful mildly interesting, and a few outright clunkers. One egregiously bad suggestion for this book's title was *Outsmarted: When Algorithms Take the Wheel*—a mismatch of title and subtitle that neither fit together nor captured the book's essence.

Second lesson: Hallucinations are not a myth, and they are not exaggerated. They're real and common enough to be problematic, particularly when you are using AI to help you write a nonfiction book. Claude was the worst culprit; a good analogy is a genius novelist with a compulsive lying habit. It could spin up a citation that looked pristine, or a zinger from a Nobel laureate that perfectly supported an argument—until you tried to verify them and realized that both were fabricated. Here are some examples of fake quotes in text suggested by Claude:

> "We segment users into over 400 different cohorts based on behavioral patterns," a former DraftKings data scientist revealed on condition of anonymity. "The AI knows exactly which messages will trigger which users. It's like having a casino host who knows your every weakness, except it's an algorithm that never sleeps."

> "Platforms face a fundamental tension," explains Einav. "They want to be trusted by buyers, but they also benefit from friction that keeps users engaged. It's like a casino—they want you to win enough to keep playing, but not so much that you cash out and leave."

> "Regulation is always fighting the last war," says Kathleen McGee, former chief of the New York Attorney General's Internet Bureau. "By the time we can prove something is deceptive, the industry has moved on to the next innovation."

All of these are great quotes, making the points I wanted to emphasize. But I could not verify any of them. These fakes were so smooth that at first I tried diligently to verify them, before I realized that researching whether they were real took longer than writing and sourcing from scratch. Picking the right model for the right task helped, but not enough to skip due diligence. My advice: If you're writing with AI, verify *everything* that carries weight—citations, references, anecdotes, quotes.

Third, humans still matter. AI can imitate an editor, a publicist, or a colleague giving feedback, but only to a certain extent. What it can't replicate is good judgment: which stories to tell, what research to foreground, how to protect a reader's time. For that, I leaned on real people—friends, colleagues, an experienced editor, and a publicist who knows how to get the word out. AI is transformative, but I felt more comfortable relying on humans to supply the ideas and direction. I recommend that other authors do the same.

Let's look at one example that highlights the value of a human editor's touch. The next two paragraphs show the same passage from my collaboration with ChatGPT—first as it appeared before editing, and then as it looked after revisions by my editor, Steven Rigolosi.

Before editor's edits:

> **Nowhere is this better illustrated than in** the rise of the gig economy. Take Uber. ... That meant a lot of **empty cruising, and other times when demand was too much to fulfill, leaving wanting customers stuck trying to** flag a cab in the rain. **But** Uber introduced surge pricing, nudging more drivers onto the road when demand was high—and fewer when it wasn't.

After:

> **An excellent example is** the rise of the gig economy. Consider Uber. ... That meant a lot of **empty cabs on warm sunny days, and too much demand to fulfill when people were trying** to flag a cab in the rain. **To fix this imbalance,** Uber introduced

> *surge pricing*, nudging more drivers onto the road when demand was high, and fewer when it wasn't.

Notice how AI tends to write with unwarranted confidence, dropping declarations like "Nowhere is this better illustrated ..." as if every example were a breakthrough. And, stylistically, AI has tells, such as its fondness for em dashes (—).

More substantively, AI Chatbots gravitate toward long, ornate sentences that sound erudite but convey little—often a side effect of uncertainty about intent. Faced with ambiguity, it errs on the side of vagueness. A human editor, by contrast, can infer intent and clarify it. For instance, my editor recognized that taxis often sit idle on sunny days, an intuitive real-world link the AI had missed, but one that immediately grounds the reader's understanding. Better-crafted prompts that more precisely convey the author's intent can reduce this problem, though some ambiguity invariably slips through.

Generative AI also falters in transitions, leaping from idea to idea. In editing the first draft of this book, the editor intervened, inserting a bridge like "To fix this imbalance," and replacing a clumsy "but" with a cleaner, more logical pivot. The revision reads as smoother, sharper, and unmistakably human. Overall, my editor noted that the AI's draft required about as many changes as a typical human-created draft, but the types of changes were slightly different.

Bottom line: AI made me quicker, but neither I nor an experienced editor was replaceable. The book reflects my ideas, structure, and judgment. For example, Chapter 10 opens with a Supreme Court case about a student sued for reselling textbooks, detours through GameStop, and lands on ReDigi, a scrappy startup trying to build a used-MP3 marketplace. Every choice—which story to highlight, which papers to cite, how to tie it all together—came from me. AI made the sentences flow more cleanly, sharply, and very quickly, *if* I provided enough detail for the AI model to confidently infer my intent and thus avoid sounding vague. Crucially, the content, structure, and intended tone are mine.

And one final note—copyright. The U.S. Copyright Office recently issued its (still evolving) position on the copyrightability of AI-generated works.[1] In the executive summary, it noted (some verbatim, some adjusted for clarity for non-lawyers):

- Copyright does not extend to purely AI-generated material, or to material with insufficient human control over the expressive elements.
- Whether human contributions to AI-generated outputs are sufficient to constitute authorship must be analyzed on a case-by-case basis.
- Human authors are entitled to copyright in their works of authorship that are perceptible in AI-generated outputs. Copyright also applies to the creative selection, coordination, and/or arrangement of material in the outputs, and to creative modifications of the outputs. In other words, copyrightable output is not something any random person could have produced with the same tool.
- The case has not been made for additional copyright or sui generis protection for AI-generated content. That is, current law does not recognize AI systems as copyright holders.

Interpretation: If you want a real chance at owning the materials you've generated with the help of AI, you need to stay firmly in the driver's seat. AI can help you with the drafting, but you must decide what remains, what's cut, and how it all comes together. A book spun out from a one-line prompt almost certainly won't qualify for copyright protection. Instead, detailed prompting, careful curation, and substantive edits tie authorship back to you. Because the legal ground is still shifting, keep an eye on the latest rules before publishing anything. Copyright is not guaranteed.

In the end, for me writing with AI was less about outsourcing creativity and more about supercharging it. The models offered sparks, polish, and speed, but the vision, direction, and accountability stayed squarely with me. That's the paradox of AI authorship: The more you lean on it, the more you need to assert your own judgment. Think of AI not as a ghostwriter but as a promising intern—sometimes brilliant and creative, sometimes careless or full of confidence about the wrong answer—who demands a steady hand on the wheel. Use it well, and you get sharper prose in less time. Use it passively, and you risk losing both clarity and credit.

Acknowledgments

Family foundation: My deepest gratitude begins with Virginia (Dr. Mom), Robert (Prof. Dad), and Derek (Dr. Brother), who built the intellectual home that made this work possible. Above all, Laurie (Wife, MD), my partner in everything and the person who keeps me grounded, makes me better every day.

Academic mentorships and colleagues: Joel Waldfogel introduced me to rigorous yet enjoyable scholarship as both mentor and collaborator. Shane Greenstein provided crucial guidance, patiently steering me back on track and enthusiastically supporting my direction. Imke Reimers has been an invaluable research partner and thoughtful collaborator. Along with Bhoomija Ranjan and Chiara Farronato, they all have generously served as insightful sounding boards throughout this journey.

Institutional support: At Wharton, Katja Seim, Alon Eizenberg, Lorin Hitt, and Eric Clemons offered generous feedback and tolerated my frequent office visits seeking advice and perspective. The Brandeis community, including Linda Bui, Steve Cecchetti, Katy Graddy, George Hall, Alice Hsiaw, Blake LeBaron, Nidhiya Menon, Aldo Musacchio, Carol Osler, Raphael Schoenle, and countless others, provided an intellectual environment where ideas flourish through countless conversations, insights, and unwavering support.

Career catalysts: Shane Greenstein, Erik Brynjolfsson, Catherine Tucker, Joshua Gans (special thanks to Joshua for his guidance on this book), Avi Goldfarb, and Hal Varian helped advance my career, offering opportunities and guidance that shaped my trajectory by opening doors and sometimes giving me the gentle push I needed to walk through them.

Digital collaborators: My AI writing companions ChatGPT, Claude, and Gemini helped me draft, revise, and brainstorm at all hours without ever requesting coffee breaks. Grok, DeepSeek, and LLaMa, perhaps you'll make the cut next time.

Editorial experts: First and foremost, I would like to thank Steven Rigolosi, whose editorial skill and insight improved this book substantially. I would also like to thank Peter Dougherty for very useful and specific advice, and Sue Ramin for sharing her perspectives. We didn't see eye to eye on everything, but their insights sharpened my thinking considerably.

Close Connections: I am grateful to my friends Cari Brown, Steve Dupree, Josh Zetlin, and Borjan Zic, and my brother Derek Shiller, for useful suggestions for improving the manuscript.

And finally, I must explicitly withhold thanks from Matt Mauro, whose relentless heckling throughout this process was of absolutely no help whatsoever. (Note: He insisted I include this statement, which tells you everything you need to know about Matt.)

Notes

2. Job Extinction Event

1. *The Economist*. "Factory Work Is Overrated. Here Are the Jobs of the Future." June 10,2025. https://www.economist.com/finance-and-economics/2025/06/10/factory-work-is-overrated-here-are-the-jobs-of-the-future.
2. U.S. Department of Agriculture, Economic Research Service. "Agriculture and Its Related Industries Provide 10.4 Percent of U.S. Employment." ERS Chart Gallery, last updated April 3, 2024. https://www.ers.usda.gov/data-products/chart-gallery/chart-detail?chartId=58282. The article explains that only 1.2% of the U.S. labor force grows food, although if you consider food retailers and waiters/servers, the fraction rises to about 10%.
3. Boston, William. "Drones, AI and Robot Pickers: Meet the Fully Autonomous Farm." *The Wall Street Journal*. July 16, 2025. https://www.wsj.com/tech/autonomous-farming-ai-95657bd1?mod=Searchresults_pos1&page=1.
4. Greicius, Anthony. "When Computers Were Human." https://www.nasa.gov/centers-and-facilities/jpl/when-computers-were-human/. Accessed Oct 14, 2025.
5. Ridenour, Louis, and George Collins. "Pulse Generators." Massachusetts Institute of Technology Radiation Laboratory Series, 1948. https://www.febo.com/pages/docs/RadLab/VOL_5_Pulse_Generators.pdf.
6. Maguire, Eleanor A., David G. Gadian, Ingrid S. Johnsrude, Catriona D. Good, John Ashburner, Richard S. J. Frackowiak, and Christopher D. Frith. "Navigation-related Structural Change in the Hippocampi of Taxi Drivers." *Proceedings of the National Academy of Sciences* 97, no. 8 (2000): 4398–4403. https://www.pnas.org/doi/abs/10.1073/pnas.070039597.
7. Manning, Benjamin S., Kehang Zhu, and John J. Horton. "Automated Social Science: Language Models as Scientist and Subjects." Working Paper. Working Paper Series. National Bureau of Economic Research, April 2024. https://doi.org/10.3386/w32381.
8. Ibid.
9. "The Future of AI: A Ubiquitous, Invisible, Smart Utility." *WSJ CIO Journal*. November 21, 2014. https://www.wsj.com/articles/BL-CIOB-5810.
10. Handa, Kunal, Alex Tamkin, Miles McCain, Saffron Huang, Esin Durmus, Sarah Heck, Jared Mueller, et al. "Which Economic Tasks Are Performed

with AI? Evidence from Millions of Claude Conversations." arXiv preprint. arXiv:2503.04761. February 11, 2025. https://arxiv.org/pdf/2503.04761.

11. Townsend, Tess. "Eric Schmidt Said ATMs Led to More Jobs for Bank Tellers. It's Not That Simple." *Vox*. May 8, 2017, 10:50 PM UTC. https://www.vox.com/2017/5/8/15584268/eric-schmidt-alphabet-automation-atm-bank-teller.
12. Autor, David, and Neil Thompson. "Expertise." NBER Working Paper w33941 (2025). https://www.nber.org/papers/w33941.
13. Humlum, Anders, and Emilie Vestergaard. "Large Language Models, Small Labor Market Effects." No. w33777. National Bureau of Economic Research, 2025. https://www.nber.org/papers/w33777.
14. Challapally, Aditya, Chris Pease, Ramesh Raskar, and Pradyumna Chari. "The GenAI Divide: State of AI in Business 2025." MIT Nanda. July 2025. https://mlq.ai/media/quarterly_decks/v0.1_State_of_AI_in_Business_2025_Report.pdf.
15. Brynjolfsson, Erik, Bharat Chandar, and Ruyu Chen. "Canaries in the Coal Mine? Six Facts About the Recent Employment Effects of Artificial Intelligence." Stanford University Working Paper. August 2025. https://digitaleconomy.stanford.edu/wp-content/uploads/2025/08/Canaries_BrynjolfssonChandarChen.pdf.
16. Singer, Natasha. "Goodbye, $165,000 Tech Jobs. Student Coders Seek Work at Chipotle." *The New York Times*. August 10, 2025. https://www.nytimes.com/2025/08/10/technology/coding-ai-jobs-students.html.
17. "OpenAI Five Defeats Dota 2 World Champions." April 15, 2019. https://openai.com/index/openai-five-defeats-dota-2-world-champions/.
18. Lotz, Avery. "AI Sycophancy: The Downside of a Digital Yes-Man." *Axios*. July 7, 2025. https://www.axios.com/2025/07/07/ai-sycophancy-chatbots-mental-health.
19. Trammell, Philip, and Anton Korinek. "Economic Growth Under Transformative AI." No. w31815. National Bureau of Economic Research, 2023. https://www.nber.org/papers/w31815.
20. Otis Elevator Company. "Our History: Elevator History Timeline." Otis.com, updated September 20, 2023. https://www.otis.com/en/us/our-company/history.
21. Freeman, Joshua B. *Working-Class New York: Life and Labor Since World War II*. The New Press, 2021. https://archive.nytimes.com/www.nytimes.com/books/first/f/freeman-newyork.html.
22. Bessen, James E. "How Computer Automation Affects Occupations: Technology, Jobs, and Skills." Boston University School of Law, Law and Economics Research Paper 15–49 (2016). https://papers.ssrn.com/sol3/papers.cfm?abstract_id=2690435.
23. Waymo. https://waymo.com/. Accessed June 29, 2025; AAA Oregon/Idaho. "For Most Americans, Autonomous Vehicles Are a No Go." https://info.

oregon.aaa.com/for-most-americans-autonomous-vehicles-are-a-no-go/. Accessed July 30, 2025.

24. Cullen, Frank. “What Is Vaudeville?” American Vaudeville Museum & UA Collections, University of Arizona. https://sites.arizona.edu/vaudeville/what-is-vaudeville-by-frank-cullen/. Accessed June 29, 2025.
25. Bloomberg. “Stand-Up Comedy Has Tripled in Size Over the Last Decade.” *Bloomberg News.* May 5, 2024. https://www.bloomberg.com/news/newsletters/2024-05-05/stand-up-comedy-has-tripled-in-size-over-the-last-decade.
26. Lohr, Steve. “Your A.I. Radiologist Will Not Be With You Soon.” *The New York Times*, May 14, 2025. Accessed July 1, 2025. https://www.nytimes.com/2025/05/14/technology/ai-jobs-radiologists-mayo-clinic.html.
27. Ibid.
28. “Radiology Rises to the No. 2 Highest Paid Specialty, Surpassing Cardiology and Plastic Surgery: Medscape.” *Radiology Business.* April 19, 2025. https://radiologybusiness.com/topics/healthcare-management/radiologist-salary/radiology-rises-no-2-highest-paid-specialty-surpassing-cardiology-and-plastic-surgery-medscape. Accessed July 1, 2025.
29. This possibility was predicted in Ajay Agrawal, Joshua Gans, and Avi Goldfarb, *Prediction Machines: The Simple Economics of Artificial Intelligence.* Harvard Business Press, 2018.

3. How Can You Avoid the Job Apocalypse?

1. For an example of which languages were supported early on, see Turovsky, Barak. “See the World in Your Language with Google Translate.” Google Blog. July 29, 2015. https://blog.google/products/translate/see-world-in-your-language-with-google/.
2. For a longer discussion, see Ajay Agrawal, Joshua Gans, and Avi Goldfarb. *Prediction Machines, Updated and Expanded: The Simple Economics of Artificial Intelligence.* Harvard Business Press, 2022.
3. Markman, Jon. “Musk: The Mountain Is 10 Billion Miles High For Safe Self-Driving” *Forbes.* January 12, 2026. https://www.forbes.com/sites/jonmarkman/2026/01/12/musk-the-mountain-is-10-billion-miles-high-for-safe-self-driving/.

   Federal Highway Administration. “Average Annual Miles per Driver by Age Group (2022).” U.S. Department of Transportation. https://www.fhwa.dot.gov/ohim/onh00/bar8.htm/. Accessed Jan 14, 2026.
4. https://epoch.ai/gradient-updates/what-went-into-training-deepseek-r1/. Accessed Jan 14, 2026.
5. One token amounts to about three quarters of a word according to OpenAI: https://platform.openai.com/tokenizer/. Accessed Jan 14, 2026.

6. Chen, M. Keith, Peter E. Rossi, Judith A. Chevalier, and Emily Oehlsen. "The Value of Flexible Work: Evidence from Uber Drivers." *Journal of Political Economy* 127, no. 6 (2019): 2735–2794. https://www.journals.uchicago.edu/doi/abs/10.1086/702171.
7. "World Record Domino Robot (100k Dominoes in 24Hours)," YouTube video, 3:56. September 8, 2021. Posted by Mark Rober. https://www.youtube.com/watch?v=8HEfIJlcFbs.
8. Alex Gray, "Kangaroos Are Confusing Self-Driving Cars," World Economic Forum (Forum Agenda), July 19, 2017. https://www.weforum.org/stories/2017/07/engineers-testing-volvo-s-driverless-technology-have-hit-a-problem-kangaroos/.
9. Robert Briel, "Companies Have Spent $16Billion on Self-Driving-Car Research," *CarandDriver*. May 1, 2019. https://www.caranddriver.com/news/a30857661/autonomous-car-self-driving-research-expensive/.
10. Humlum, Anders, and Pernille Plato. *Reskilling and Resilience*. No. w34095. National Bureau of Economic Research, 2025. https://www.nber.org/system/files/working_papers/w34095/w34095.pdf.

4. The Education Rip-Off

1. McCormack, Laura K.C. "RomanEducation." *World History Encyclopedia*. April 24, 2023. https://www.worldhistory.org/article/2224/roman-education/.
2. NCES. "Pupil-Teacher Ratio in Public and Private Elementary and Secondary Schools in the United States from 1955 to 2031. " Chart. March 31, 2023. *Statista*. https://www.statista.com/statistics/185021/pupil-teacher-ratio-in-elementary-and-secondary-schools-since-1955/. Accessed October 03, 2025.
3. Esaki-Smith, Anna. "Yearly Cost of Going to Wellesley College Exceeds $100,000." *Forbes*. April 3, 2025. https://www.forbes.com/sites/annaesakismith/2025/04/03/yearly-cost-of-going-to-wellesley-college-exceeds-100000/.
4. Hanson, Melanie. "U.S. Public Education Spending Statistics" *EducationData.org*. February 8, 2025. Accessed October 3, 2025. https://educationdata.org/public-education-spending-statistics.
5. Bloom, Benjamin S. "The 2 Sigma Problem: The Search for Methods of Group Instruction as Effective as One-To-One Tutoring." *Educational Researcher* 13, no. 6 (1984): 4–16. https://journals.sagepub.com/doi/abs/10.3102/0013189X013006004.
6. Nickow, Andre, Philip Oreopoulos, and Vincent Quan. "The Impressive Effects of Tutoring on PreK-12 Learning: A Systematic Review and Meta-Analysis of the Experimental Evidence." July 2020. https://doi.org/10.3386/w27476.

7. De Simone, Martín E., Federico Tiberti, Wuraola Mosuro, Federico Manolio, Maria Barron, and Eliot Dikoru. “From Chalkboards to Chatbots: Transforming Learning in Nigeria, One Prompt at a Time.” Education for Global Development, World Bank Blogs. January 9, 2025. https://blogs.worldbank.org/en/education/From-chalkboards-to-chatbots-Transforming-learning-in-Nigeria.
8. Ibid.
9. Anthropic. “Rising Academies’ Chatbot Tutors Reach 150,000+ Students Across Sub-Saharan Africa with Claude.” *Anthropic Customers: Case Studies*. March 27, 2025. https://www.anthropic.com/customers/rising-academies.
10. *The Economist*. “Can AI Be Trusted in Schools?” May 30, 2025. https://www.economist.com/graphic-detail/2025/05/30/can-ai-be-trusted-in-schools.
11. Alpha’s webpage. https://alpha.school/the-program/. Accessed August 27, 2025.
12. Rothwell, Jonathan. “Teens Spend Average of 4.8 Hours on Social Media Per Day.” *Gallup News*. October 13, 2023. https://news.gallup.com/poll/512576/teens-spend-average-hours-social-media-per-day.aspx.
13. Morrison, Briana B., and Betsy DiSalvo. “Khan Academy Gamifies Computer Science.” In Proceedings of the 45th ACM Technical Symposium on Computer Science Education (SIGCSE '14). Association for Computing Machinery, New York, NY, USA, 39–44. 2014. https://doi.org/10.1145/2538862.2538946.

5. Total Inequality

1. Dickmanns, Ernst. An Interview Conducted by Peter Asaro for the IEEE History Center, June 21, 2010. Interview #681 for Indiana University and the IEEE History Center, Institute of Electrical and Electronics Engineers, Inc. https://ethw.org/Oral-History:Ernst_Dickmanns#Entering_the_Aerospace_Engineering_Field.
2. Reimers, Imke, and Benjamin Reed Shiller. “Will New Driving Technologies Change the Value of Public Transportation Investments?” CESifo Working Paper, Munich: Ifo Institute, 2025. https://www.ifo.de/en/cesifo/publications/2025/working-paper/will-new-driving-technologies-change-value-public-transportation.
3. Duranton, Gilles, and Matthew A. Turner. "The Fundamental Law of Road Congestion: Evidence from US Cities." *American Economic Review* 101, no. 6 (2011): 2616–2652. https://www.aeaweb.org/articles?id=10.1257/aer.101.6.2616.
4. Acemoglu, Daron, and Pascual Restrepo. "Tasks, Automation, and the Rise in US Wage Inequality." *Econometrica* 90, no. 5 (2022): 1973–2016. https://onlinelibrary.wiley.com/doi/full/10.3982/ECTA19815.

5. Brynjolfsson, Erik. "The Turing Trap: The Promise and Peril of Human-Like Artificial Intelligence." https://digitaleconomy.stanford.edu/news/the-turing-trap-the-promise-peril-of-human-like-artificial-intelligence/.
6. Agrawal, Ajay, Joshua Gans, and Avi Goldfarb. *Prediction Machines: The Simple Economics of Artificial Intelligence*. Harvard Business Press, 2018.
7. Liang, Yang, Joseph J. Sabia, and Dhaval M. Dave. "Robots and Crime." Working Paper. Working Paper Series. National Bureau of Economic Research, March 2025. https://doi.org/10.3386/w33603.
8. Fajnzylber, Pablo, Daniel Lederman, and Norman Loayza. "Inequality and Violent Crime." *Journal of Law and Economics* 45, no. 1 (2002): 1–40. https://doi.org/10.1086/338347.
9. Ha, Anthony. "Authors Call on Publishers to Limit Their Use of AI," *TechCrunch*. June 28, 2025. https://techcrunch.com/2025/06/28/authors-call-on-publishers-to-limit-their-use-of-ai.
10. Ballotpedia, "Presidential Statewide Margins of Victory of Five Percentage Points or Fewer, 1948–2024." https://ballotpedia.org/Presidential_statewide_margins_of_victory_of_five_percentage_points_or_fewer%2C_1948-2024.
11. "Luddite." *Encyclopædia Britannica*, updated most recently by The Editors of Encyclopædia Britannica, originally published in 1998. https://www.britannica.com/event/Luddite. Accessed June 29, 2025.
12. Vivalt, Eva, Elizabeth Rhodes, Alexander W. Bartik, David E. Broockman, Patrick Krause, and Sarah Miller. "The Employment Effects of a Guaranteed Income: Experimental Evidence from Two US States." No. w32719. National Bureau of Economic Research, 2024. https://www.nber.org/papers/w32719.
13. Bonn, Tess. "Sanders Criticizes Yang's Universal Basic Income Proposal: 'People Want to Work.'" *The Hill*. August 27, 2019. https://thehill.com/hilltv/rising/458951-sanders-yangs-universal-basic-income-proposal-not-a-solution-to-impact-of-automation/.
14. To be clear, violence is never justified, and it isn't a solution. But it will happen sometimes, and more often as inequality increases.

6. Is AI Going to Kill Us All (Eventually)?

1. Boutin, Paul. "Where Are They?" *MIT Technology Review*. April 22, 2008. https://www.technologyreview.com/2008/04/22/220999/where-are-they/.
2. Bendett, Samuel, and David Kirichenko. "Battlefield Drones and the Accelerating Autonomous Arms Race in Ukraine." Modern War Institute, U.S. Military Academy. January 10, 2025. https://mwi.westpoint.edu/battlefield-drones-and-the-accelerating-autonomous-arms-race-in-ukraine/.
3. Hambling, David. "Ukraine Drone Carriers Launch First-Long-Range Autonomous Strikes." *Forbes*. May 26, 2025. https://www.forbes.com/sites/

davidhambling/2025/05/26/ukraine-drone-carriers-launch-first-long-range-autonomous-strikes/.

4. CRISPR is an abbreviation for Clustered Regularly Interspaced Short Palindromic Repeats. In practice, it allows users to modify DNA of humans, germs, animals, and so on.
5. Bostrom, Nick. *Superintelligence: Paths, Dangers, Strategies*. Oxford: Oxford University Press, 2014.
6. CNN Business. "Sam Altman Warns AI Could Kill Us All. But He Still Wants the World to Use It." CNN Business. October 31, 2023. https://www.cnn.com/2023/10/31/tech/sam-altman-ai-risk-taker.
7. Axios. "Behind the Curtain: What If They're Right?" Axios AM. June 16, 2025. https://www.axios.com/2025/06/16/ai-doom-risk-anthropic-openai-google.
8. "What If AI Made the World's Economic Growth Explode?" *The Economist*. July 24, 2025.
   https://www.economist.com/briefing/2025/07/24/what-if-ai-made-the-worlds-economic-growth-explode.
9. Jones, Charles I. "The AI Dilemma: Growth versus Existential Risk." *American Economic Review: Insights* 6, no. 4 (December 1, 2024): 575–90. https://doi.org/10.1257/aeri.20230570.

7. Data Nudity

1. Associated Press. "China Takes DIY Approach to Mountain Greenery." *The Guardian*. February 14, 2007. https://www.theguardian.com/world/2007/feb/14/china.
2. Zhao, Xuan, and Li Rongde. "What Bad Air? Hunan Officials Use Mist Cannons to Fool Pollution Meters." Caixin Global. February 2, 2018. https://www.caixinglobal.com/2018-02-02/what-bad-air-hunan-officials-use-mist-cannons-to-fool-pollution-meters-101206784.html.
3. Greenstone, Michael, Guojun He, Ruixue Jia, and Tong Liu. "Can Technology Solve the Principal-Agent Problem? Evidence from China's War on Air Pollution." *American Economic Review: Insights* 4, no. 1 (March 2022): 54–70. https://doi.org/10.1257/aeri.20200373.
4. Ibid.
5. Reimers, Imke, and Benjamin R. Shiller. "The Impacts of Telematics on Competition and Consumer Behavior in Insurance." *The Journal of Law and Economics* 62, no. 4 (November 2019): 613–32. https://doi.org/10.1086/705119.
6. J.D. Power, "Auto Insurance Customer Satisfaction Plummets as Rates Continue to Surge, J.D. Power Finds." June 13, 2023. https://www.jdpower.com/business/press-releases/2023-us-auto-insurance-study. Accessed October 7, 2025.

7. Hill, Kashmir. "Automakers Are Sharing Consumers' Driving Behavior with Insurance Companies." *The New York Times*. March 11, 2024. https://www.nytimes.com/2024/03/11/technology/carmakers-driver-tracking-insurance.html?ref=oembed.
8. Hawkins, Andre. "GM Banned from Selling Your Driving Data for Five Years." *The Verge*. January 16, 2025. https://www.theverge.com/2025/1/16/24345470/gm-banned-selling-driving-data-insurance-ftc.
9. https://www.tesla.com/support/insurance/tesla-real-time-insurance?utm_source=chatgpt.com.
10. Handel, Ben, Igal Hendel, and Michael D. Whinston. "Equilibria in health exchanges: Adverse selection versus reclassification risk." *Econometrica* 83, no. 4 (July 2015): 1261-1313. https://doi.org/10.3982/ECTA12480.
11. Lecher, Collin. "How Amazon Automatically Tracks and Fires Warehouse Workers for 'Productivity.'" *The Verge*. April 25, 2019. https://www.theverge.com/2019/4/25/18516004/amazon-warehouse-fulfillment-centers-productivity-firing-terminations?utm_source=chatgpt.com.
12. Adolphus, Emell. "Elon Musk Puts His AI Company's Employees Under Surveillance." *The Daily Beast*. July 13, 2025. https://www.thedailybeast.com/elon-musk-puts-his-ai-companys-employees-under-surveillance/?utm_source=chatgpt.com.
13. Bangalore, Sanvi. "The Jiggle Is Up: Bosses Bust Workers Who Fake Computer Activity." *The Wall Street Journal*. July 2, 2024. https://www.wsj.com/lifestyle/workplace/the-jiggle-is-up-bosses-bust-workers-who-fake-computer-activity-b6374f22?utm_source=chatgpt.com.

8. INVOLUNTARY EXPOSURE

1. Goldfarb, Avi, and Catherine E. Tucker, eds. *The Economics of Privacy*. Chicago: University of Chicago Press, 2024.
2. Drew Harwell, "Is Your Pregnancy App Sharing Your Intimate Data with Your Boss?" *The Washington Post*. April 10, 2019. https://www.washingtonpost.com/technology/2019/04/10/tracking-your-pregnancy-an-app-may-be-more-public-than-you-think/.
3. Senate Committee on Commerce, Science, and Transportation. *Marketing of Consumer Information*. C-SPAN. December 18, 2013. https://www.c-span.org/program/senate-committee/marketing-of-consumer-information/333106.
4. Kim, Joanne. "Data Brokers and the Sale of Americans' Mental Health Data: The Exchange of Our Most Sensitive Data and What It Means for Personal Privacy." Durham, NC: Duke University Tech Policy Lab, 2025. https://techpolicy.sanford.duke.edu/data-brokers-and-the-sale-of-americans-mental-health-data/.

5. "OpenAI CEO Sam Altman on GPT-5: We've Built an 'Integrated Single Experience.'" *Squawk Box*. CNBC Television. August 8, 2025. https://www.youtube.com/watch?v=xx_5ADX21-4.
6. Pazur, Barabard. "What Are Deepfakes? Everything You Need to Know About These AI Image and Video Forgeries." *CNET.com*. May 21, 2025. https://www.cnet.com/tech/services-and-software/what-are-deepfakes-everything-to-know-about-these-ai-image-and-video-forgeries/.
7. Ramer, Holly. "Consultant Behind AI-Generated Robocalls Mimicking Biden Goes on Trial in New Hampshire." AP News. June 5, 2025. https://www.apnews.com/article/ai-robocalls-biden-kramer-new-hampshire-02de549e5d82bce4cd7622b8bef2d587.
8. Allyn, Bobby. "Deepfake Video of Zelenskyy Could Be 'Tip of the Iceberg' in Info War, Experts Warn." NPR, March 16, 2022. https://www.npr.org/2022/03/16/1087062648/deepfake-video-zelenskyy-experts-war-manipulation-ukraine-russia.
9. Goldman, Sharon. "How a Deepfake of Marco Rubio Exposed the Alarming Ease of AI Voice Scams." *Fortune*. July 10, 2025. https://www.fortune.com/2025/07/10/deepfake-marco-rubio-ai-voice-scams/.
10. Elsner, Mark, Grace Atkinson, and Saadia Zahidi. *Global Risks Report 2025*. Geneva: World Economic Forum, January 15, 2025. https://reports.weforum.org/docs/WEF_Global_Risks_Report_2025.pdf.
11. Dugas, Andrea Freyer, Mehdi Jalalpour, Yulia Gel, Scott Levin, Fred Torcaso, Takeru Igusa, and Richard E. Rothman. "Influenza Forecasting with Google Flu Trends." *PLOS One* 8, no. 2 (2013): e56176. https://doi.org/10.1371/journal.pone.0056176.
12. Mao, Jialin, Michael Matheny, Kim G. Smolderen, Carlos Mena-Hurtado, Art Sedrakyan, and Philip Goodney. "Combining Electronic Health Records Data from a Clinical Research Network with Registry Data to Examine Long-Term Outcomes of Interventions and Devices: An Observational Cohort Study." *BMJ Open* 14, no. 9 (September 26, 2024): e085806. https://doi.org/10.1136/bmjopen-2024-085806.
13. Science Writers at Clue. "How Tracking Your Cycle Advances Female Health." Clue. https://helloclue.com/articles/about-clue/scientific-research-at-clue. Accessed June 28, 2025.
14. Shogren, Elizabeth. "White House to Seek Genetic Test Safeguards." *Los Angeles Times*. January 20, 1998. https://www.latimes.com/archives/la-xpm-1998-jan-20-mn-10164-story.html.
15. Hayden, Sommer, Sarah Mange, Debra Duquette, Nancie Petrucelli, Victoria M. Raymond, and BRCA Clinical Network Partners. "Large, Prospective Analysis of the Reasons Patients Do Not Pursue BRCA Genetic Testing Following Genetic Counseling." *Journal of Genetic Counseling* 26, no. 4 (2017): 859–865. https://link.springer.com/article/10.1007/s10897-016-0064-5.

16. American Psychological Association. "Protecting Your Privacy: Understanding Confidentiality in Psychotherapy." https://www.apa.org/topics/psychotherapy/confidentiality. Accessed June 28, 2025.
17. Joinson, Adam N. "Self-Disclosure in Computer-Mediated Communication: The Role of Self-Awareness and Visual Anonymity." *European Journal of Social Psychology* 31, no. 2 (2001): 177–192. https://doi.org/10.1002/ejsp.36.
18. Verve. "More Control Drives More Data Sharing: Surprising In-App Privacy Trends Revealed by Verve." Verve Press. https://verve.com/press/more-control-drives-more-data-sharing-surprising-in-app-privacy-trends-revealed-by-verve/. Accessed June 28, 2025.
19. Newman, Lily Hay. "The Privacy Battle to Save Google from Itself." *Wired*. Nov 1, 2010. https://www.wired.com/story/google-privacy-data/.
20. Walsh, Dylan. "GDPR Reduced Firms' Data and Computation Use." MIT Sloan Idea Made to Matter, September 10, 2024. https://mitsloan.mit.edu/ideas-made-to-matter/gdpr-reduced-firms-data-and-computation-use.
21. Janssen, Rebecca, Reinhold Kesler, Michael E. Kummer, and Joel Waldfogel. "GDPR and the Lost Generation of Innovative Apps." No. w30028. National Bureau of Economic Research, 2022. https://www.nber.org/papers/w30028.

9. Psychic Pricing

1. McCormack, Laura K.C. "John Wanamaker Makes a Sale." *Cosmos*, 2021. https://cosmosmagazine.com/people/john-wanamaker-makes-a-sale/.
2. "Amazon.com's Variable Pricing Draws Ire." *Chicago Tribune*. October 9, 2000. https://www.chicagotribune.com/2000/10/09/amazoncoms-variable-pricing-draws-ire/.
3. Kahneman, Daniel, Jack L. Knetsch, and Richard Thaler. "Fairness as a Constraint on Profit Seeking: Entitlements in the Market." *The American Economic Review* 76, no. 4 (1986): 728–41. https://www.jstor.org/stable/1806070.
4. Shiller, Benjamin Reed. "Approximating Purchase Propensities and Reservation Prices from Broad Consumer Tracking." *International Economic Review* 61, no. 2 (2020): 847–870. https://onlinelibrary.wiley.com/doi/10.1111/iere.12442.
5. Hannak, Aniko, Gary Soeller, David Lazer, Alan Mislove, and Christo Wilson. "Measuring Price Discrimination and Steering on e-Commerce Web Sites." In Proceedings of the 2014 Conference on Internet Measurement Conference, pp. 305–318. 2014. https://dl.acm.org/doi/abs/10.1145/2663716.2663744;

   Mikians, Jakub, László Gyarmati, Vijay Erramilli, and Nikolaos Laoutaris. "Detecting Price and Search Discrimination on the Internet." In Proceedings of the 11th ACM Workshop on Hot Topics in Networks, pp. 79–84. 2012. https://dl.acm.org/doi/abs/10.1145/2390231.2390245.

6. Goli, Ali. *Personalized Versioning: Product Strategies Constructed from Experiments on Pandora.* Ph.D. dissertation. The University of Chicago, 2020. https://www.proquest.com/docview/2449494718?pq-origsite=gscholar&fromopenview=true&sourcetype=Dissertations%20&%20Theses.
7. Shiller, Benjamin. "Inconspicuous Personalized Pricing." *Journal of Industrial Economics*. Forthcoming.
8. Ivanova, Irina. "Delta Moves Toward Eliminating Set Prices in Favor of AI That Determines How Much You Personally Will Pay for a Ticket." *Fortune.* July 16, 2025. https://fortune.com/2025/07/16/delta-moves-toward-eliminating-set-prices-in-favor-of-ai-that-determines-how-much-you-personally-will-pay-for-a-ticket/.
9. Ibid.
10. Shove, G. F. "Review of *The Economics of Imperfect Competition* by Joan Robinson." *The Economic Journal*, 43, no. 172 (1933): 657–661. https://doi.org/10.2307/2224510; Bergemann, Dirk, Benjamin Brooks, and Stephen Morris. "The Limits of Price Discrimination." *American Economic Review* 105, no. 3 (2015): 921–957. https://www.aeaweb.org/articles?id=10.1257/aer.20130848.
11. Thisse, Jacques-Francois, and Xavier Vives. "On The Strategic Choice of Spatial Price Policy." *The American Economic Review* 78, no. 1 (1988): 122–137.

10. Digital Serfdom

1. Kari, Doug. "How an eBay Bookseller Defeated a Publishing Giant at the Supreme Court." *Ars Technica*. November 25, 2014. https://arstechnica.com/tech-policy/2014/11/how-an-ebay-bookseller-defeated-a-publishing-giant-at-the-supreme-court/.
2. Ibid.
3. Ibid.
4. Ibid.
5. Ibid.
6. Drug companies still price their products differently by country, sidestepping the patent-law version of the first-sale doctrine, called "exhaustion," by invoking consumer safety rules that apply specifically to pharmaceuticals.
7. GameStop Corp. Annual Report 2005. Grapevine, TX: GameStop Corp. 2005. https://www.annualreports.com/HostedData/AnnualReportArchive/g/NYSE_GME_2005.pdf.
8. Statista. "Net Sales of GameStop Worldwide from 2010 to 2012 (in Billion U.S. Dollars)." *Statista.* https://www.statista.com/statistics/284523/net-sales-of-gamestop-worldwide-2010-2012/. Accessed June 29, 2025.

9. Ars Technica. "RIAA Wants ReDigi Out of the Business of Selling 'Used' iTunes Tracks." *Ars Technica*. November 15, 2011. https://arstechnica.com/tech-policy/2011/11/riaa-wants-redigi-out-of-the-business-of-selling-used-itunes-tracks/.
10. Physical copies eventually degrade. Well-crafted hardcover books typically last over 100 checkouts. Rebseman, Werner. "Superior Materials Used in Library Binding Make the Difference!" *The Library Scene* 22, no. 3 (2003). https://bindery.berkeley.edu/sites/default/files/LibraryBindingDifference.pdf.
11. Roose, Robert. "The True Cost of eBooks and Audiobooks for Libraries." Spokane Public Library Blog. January 14, 2025. https://www.spokanelibrary.org/the-true-cost-of-ebooks-and-audiobooks-for-libraries/.

11. The Savvy Consumer Delusion

1. Ellison, Glenn, and Sara Fisher Ellison. "Match Quality, Search, and the Internet Market for Used Books." No. w24197. National Bureau of Economic Research. 2018. https://www.nber.org/papers/w24197.
2. Anderson, Simon P., and Joshua S. Gans, "Platform Siphoning: Ad-Avoidance and Media Content," *American Economic Journal: Microeconomics* 3, no. 4 (November 2011): 1–34. https://doi.org/10.1257/mic.3.4.1.
3. "How Many People Use Ad Blockers?" *Marketing Charts*. May 23, 2024. https://www.marketingcharts.com/advertising-trends-233103.
4. Ibid.
5. Hayes, Dade. "NBCUniversal Unveils Advertising Initiatives Including 'Content Quality Index,' 29 New Measurement Partners, Expanded 'In-Scene' Push." *Deadline*. February 8, 2023. https://deadline.com/2023/02/nbcuniversal-one23-advertising-measurement-streaming-1235253488/.
6. Prince, Matthew. "Content Independence Day: No AI Crawl Without Compensation!" *The Cloudflare Blog*. July 1, 2025. https://blog.cloudflare.com/content-independence-day-no-ai-crawl-without-compensation/.

12. Confuse & Conquer?

1. Milden, Dashia. "Don't Let Sneaky Subscriptions Ruin Your Budget. Americans Spend More Than $1,000 a Year on These Services, CNET Survey Finds." June 18, 2025. https://www.cnet.com/personal-finance/subscription-survey-2025/.
2. Ellison, Glenn, and Sara Fisher Ellison. "Search, Obfuscation, and Price Elasticities on the Internet." *Econometrica* 77, no. 2 (2009): 427–452. https://onlinelibrary.wiley.com/doi/abs/10.3982/ECTA5708.
3. "Ryanair Toilet Charge Is No Joke, Insists O'Leary." *The Guardian*. March 5, 2009. https://www.theguardian.com/business/2009/mar/05/ryanair-toilet-charge.

4. Topham, Gwyn. "Many Ryanair Flights Could be Free in a Decade, Says Its Chief." *The Guardian*. November 22, 2016. https://www.theguardian.com/business/2016/nov/22/ryanair-flights-free-michael-oleary-airports.
5. Shampanier, Kristina, Nina Mazar, and Dan Ariely. "Zero as a Special Price: The True Value of Free Products." *Marketing Science* 26, no. 6 (2007): 742–757. https://pubsonline.informs.org/doi/abs/10.1287/mksc.1060.0254.
6. Tuttle, Brad. "You Probably Spent 13 Hours on Hold Last Year." *Time*. January 24, 2013. https://business.time.com/2013/01/24/you-probably-spent-13-hours-on-hold-last-year/.
7. Beshears, John, James J. Choi, David Laibson, and Brigitte C. Madrian. "The Importance of Default Options for Retirement Saving Outcomes: Evidence from the United States." In *Social Security Policy in a Changing Environment*, pp. 167–195. University of Chicago Press, 2009. https://www.nber.org/system/files/chapters/c4539/c4539.pdf.
8. Della Vigna, Stefano, and Ulrike Malmendier. "Paying Not to Go to the Gym." *American Economic Review* 96, no. 3 (2006): 694–719. https://www.aeaweb.org/articles?id=10.1257/aer.96.3.694.
9. Quick, Jason. "'I Literally Can't Stop.' The Descent of a Modern Sports Fan." *The New York Times*. October 15, 2024. https://www.nytimes.com/athletic/5777632/2024/10/14/sports-betting-addiction-problem-fans/.
10. Rocket Companies. "Rocket Companies to Acquire Truebill, Adding Rapidly Expanding Financial Empowerment FinTech to the Rocket Platform." https://ir.rocketcompanies.com/news-and-events/press-releases/press-release-details/2021/Rocket-Companies-to-Acquire-Truebill-Adding-Rapidly-Expanding-Financial-Empowerment-FinTech-to-the-Rocket-Platform/default.aspx
11. Amazon 2024 Annual Report. https://s2.q4cdn.com/299287126/files/doc_financials/2025/ar/Amazon-2024-Annual-Report.pdf.
12. Dinerstein, Michael, Liran Einav, Jonathan Levin, and Neel Sundaresan. "Consumer Price Search and Platform Design in Internet Commerce." *American Economic Review* 108, no. 7 (2018): 1820–1859. https://www.aeaweb.org/articles?id=10.1257/aer.20171218.

13. The New Economics of Deception

1. Rogers, Sam. "International Scammers Steal Over $1 Trillion in 12 Months in Global State of Scams Report 2024." Global Anti-Scam Alliance. https://www.gasa.org/post/global-state-of-scams-report-2024-1-trillion-stolen-in-12-months-gasa-feedzai. Accessed July 29, 2025.
2. O'Neill, Aaron. "The 20 Countries with the Largest Gross Domestic Product (GDP) in 2025." *Statista*. https://www.statista.com/statistics/268173/countries-with-the-largest-gross-domestic-product-gdp/.

3. "New FTC Data Show a Big Jump in Reported Losses to Fraud to $12.5 Billion in 2024." Federal Trade Commission Press Release. https://www.ftc.gov/news-events/news/press-releases/2025/03/new-ftc-data-show-big-jump-reported-losses-fraud-125-billion-2024.
4. "How Dangerous Is Lightning?" *The National Weather Service.* https://www.weather.gov/safety/lightning-odds. Accessed Oct 9, 2025.
5. "Cambodia: Government Allows Slavery and Torture to Flourish Inside Hellish Scamming Compounds." Amnesty International. June 26, 2025. https://www.amnesty.org/en/latest/news/2025/06/cambodia-government-allows-slavery-torture-flourish-inside-scamming-compounds/.
6. Nakashima, Ellen. "Feds Recover More Than $2 Million in Ransomware Payments from Colonial Pipeline Hackers." *The Washington Post.* June 7, 2021. https://www.washingtonpost.com/business/2021/06/07/colonial-pipeline-ransomware-payment-recovered/.

14. Should Most e-Books Be Free?

1. Aguiar, Luis, Imke Reimers, and Joel Waldfogel. "Platforms and the Transformation of the Content Industries." *Journal of Economics and Management Strategy* 33, no. 2 (2024): 317–326. https://doi.org/10.1111/jems.12519.
2. Discovery. "The 34 Best Selling Self-Published Books of the Past 100+ Years." https://reedsy.com/discovery/blog/best-selling-self-published-books. Accessed June 28, 2025.
3. Peukert, Christian, and Imke Reimers. "Digitization, Prediction, and Market Efficiency: Evidence from Book Publishing Deals." *Management Science* 68, no. 9 (2022): 6907–6924. https://pubsonline.informs.org/doi/abs/10.1287/mnsc.2021.4237.
4. Wakabayashi, Daisuke, and Spencer E. Ante. "Mobile Game Fight Goes Global." *The Wall Street Journal.* June 14, 2012. http://online.wsj.com/article/SB10001424052702303410404577463990080511160.html.
5. Pollstar. "2016 Year End Business Analysis." January 2017. https://data.pollstar.com/chart/2017/01/2016YearEndBusinessAnalysis_342.pdf.
6. Oberholzer-Gee, Felix, and Koleman Strumpf. "The Effect of File Sharing on Record Sales: An Empirical Analysis." *Journal of Political Economy* 115, no. 1 (2007): 1-42. https://www.jstor.org/stable/10.1086/511995.
7. Hendricks, Ken, and Alan Sorensen. "Information and the Skewness of Music Sales." *Journal of Political Economy* 117, no. 2 (April 2009): 324–69. https://doi.org/10.1086/599283.
8. Shiller, Robert J. *Irrational Exuberance.* Princeton University Press, 2001.
9. Noy, Shakked, and Whitney Zhang. "Experimental Evidence on the Productivity Effects of Generative Artificial Intelligence." *Science* 381, no. 6654 (July 14, 2023): 187–192. https://doi.org/10.1126/science.adh2586.

15. THE "STUPID" GENIUS?

1. De Cosmo, Leonardo. "Google Engineer Claims AI Chatbot Is Sentient: Why That Matters." *Scientific American*. July 12, 2022. https://www.scientificamerican.com/article/google-engineer-claims-ai-chatbot-is-sentient-why-that-matters/.
2. Lee, Timothy B. "How a Stubborn Computer Scientist Accidentally Launched the Deep Learning Boom." *Ars Technica*. November 14, 2024. https://arstechnica.com/ai/2024/11/how-a-stubborn-computer-scientist-accidentally-launched-the-deep-learning-boom/.
3. Sutton, Richard. "The Bitter Lesson." *Incomplete Ideas* (blog) 13, no. 1 (2019): 38. https://heartyhaven.github.io/files/bitter_lesson.pdf.
4. Lee, Timothy B. "A Jargon-Free Explanation of How AI Large Language Models Work." *Ars Technica*. July 31, 2023. https://arstechnica.com/science/2023/07/a-jargon-free-explanation-of-how-ai-large-language-models-work/.
5. Epoch AI. "Will We Run Out of Data? Limits of LLM Scaling Based on Human-Generated Data." Epoch AI Blog. June 18, 2024. https://epoch.ai/blog/will-we-run-out-of-data-limits-of-llm-scaling-based-on-human-generated-data.

AFTERWORD

1. United States Copyright Office. "Copyright and Artificial Intelligence: Part 2: Copyrightability." A Report of the Register of Copyrights. January 2025. https://www.copyright.gov/ai/Copyright-and-Artificial-Intelligence-Part-2-Copyrightability-Report.pdf.

Index

A

B

C

D

E

F

G

H

I

J

K

L

M

N

O

P

R

S

T

U

W

Y

Z

www.ingramcontent.com/pod-product-compliance
Lightning Source LLC
Chambersburg PA
CBHW071553210326
41597CB00019B/3218

* 9 7 9 8 9 9 3 9 3 1 2 1 0 *